PAKISTAN'S NUCLEAR FUTURE

Reining in the Risk

Attention Authors

Manas Publications is fighting a war to tell the world that India can win the battle not only by the bullet but also by pen. We are converting fighters into writers. There is no dearth of intellectuals in our country, however, their knowledge is confined to them only. No sincere effort has been made by any other publisher to channelise their knowledge, which is why the talent of every intellectual remains latent. An author always presents the raw material in the shape of a manuscript and it is the job of the publisher to convert it to a finished product. Manas has been continually motivating intellectuals and publishing their manuscripts for more than two decades and would like to publish your manuscript too. If you or your colleagues have a manuscript ready for publication or are working on it, contact us with a detailed synopsis and list of contents. Manas will also suggest you a suitable title for writing related to your area of expertise. Manas is known for its publishing quality. We take utmost care in publishing and also give wide publicity to the book and its author through our national and international distribution network.

PAKISTAN'S NUCLEAR FUTURE

Reining in the Risk

Editor

Henry D Sokolski

Manas Publications

(We Convert Fighters into Writers)

New Delhi-110002 (INDIA)

MANAS PUBLICATIONS
(Publishers, Library Suppliers, Importers & Exporters)
4402/5-A, Ansari Road
(Opp. HDFC Bank)
Darya Ganj, New Delhi-110 002 (INDIA)
Off. ✆ 23260783, 23265523, Res. ✆ 23842660
Fax: 011-23272766
E-mail: manaspublications@vsnl.com
manaspublications@gmail.com
www.manaspublications.in

ISBN 978-81-7049-389-1
Rs. 995

Typeset at
Manas Publications

Printed in India at
RK Print Services, Delhi
and published by Mrs Suman Lata for
Manas Publications, 4402/5-A, Ansari Road
(Opp. HDFC Bank),
Darya Ganj, New Delhi-110 002 (INDIA)

Contents

Preface

Unfortunately, a nuclear terrorist act is only one—and hardly the most probable—of several frightening security threats Pakistan now faces or poses. We know that traditional acts of terrorism and conventional military crises in Southwest Asia have nearly escalated into wars and, more recently, even threatened Indian and Pakistani nuclear use. Certainly, the war jitters that attended the recent terrorist attacks against Mumbai highlighted the nexus between conventional terrorism and war. For several weeks, the key worry in Washington was that India and Pakistan might not be able to avoid war. Similar concerns were raised during the Kargil crisis in 1999, and the Indo-Pakistani conventional military tensions that arose in 2001 and 2002—crises that most analysts (including those who contributed to this volume) believe could have escalated into nuclear conflicts. The intent of this book is to conduct a significant evaluation of these threats. These analyses offer a window into what is possible and why Pakistani nuclear terrorism is best seen as a lesser included threat to war, and terrorism more generally. Could the United States do more with Pakistan to secure Pakistan's nuclear weapons holdings against possible seizure? It is unclear. This book argues that rather than distracting our policy leaders from taking the steps needed to reduce the threats of nuclear war, we would do well to view our worst terrorist nightmares for what they are: Subordinate threats that will be limited best if the risk of nuclear war is reduced and contained.

Introduction

Pakistan's Nuclear Plans:
What's Worrisome, What's Avertable?

—Henry D Sokolski

With any attempt to assess security threats, there is a natural tendency to focus first on the worst. Consider the most recent appraisals of Pakistan's nuclear programme. Normally, the risk of war between Pakistan and India and possible nuclear escalation would be bad enough. Now, however, most American security experts are riveted on the frightening possibility of Pakistani nuclear weapons capabilities falling into the hands of terrorists intent on attacking the United States.[1]

Presented with the horrific implications of such an attack, the American public and media increasingly have come to view nearly all Pakistani security issues through this lens. Public airing of these fears, in turn, appear now to be influencing terrorist operations in Pakistan.[2]

Unfortunately, a nuclear terrorist act is only one—and hardly the most probable—of several frightening security threats Pakistan now faces or poses. We know that traditional acts of terrorism and conventional military crises in South West Asia have nearly escalated into wars and, more recently, even threatened Indian and Pakistani nuclear use.

Certainly, the war jitters that attended the recent terrorist attacks against Mumbai highlighted the nexus between conventional terrorism and war. For several weeks, the key worry in Washington was that India and Pakistan might not be able to avoid war.[3] Similar concerns were raised during the Kargil crisis in 1999 and the Indo-Pakistani conventional military tensions that arose in 2001 and 2002—crises that most analysts (including those who contributed to this volume) believe could have escalated into nuclear conflicts.

This book is meant to take as long a look at these threats as possible. Its companion volume, *Worries Beyond War,* published last year, focused on the challenges of Pakistani nuclear terrorism.[4] These analyses offer a window into what is possible and why Pakistani nuclear terrorism is best seen as a lesserincluded threat to war, and terrorism more generally.

Could the United States do more with Pakistan to secure Pakistan's nuclear weapons holdings against possible seizure? It is unclear. News reports indicate that the United States has already spent $100 million toward this end. What this money has bought, however, has only been intimated. We know that permissive action link (PAL) technology that could severely complicate unauthorized use of existing Pakistani weapons (and would require Pakistan to reveal critical weapons design specifics to the United States that might conceivably allow the United States to remotely "kill" Pakistani weapons) was not shared. Security surveillance cameras and related training, on the other hand, probably were.[5]

Meanwhile, the Pakistani military—anxious to ward off possible preemptive attacks against its nuclear weapons assets—remains deeply suspicious of the United States or any other foreign power trying to learn more about the number, location, and physical security of Pakistan's nuclear weapons holdings.[6] Conducting secret, bilateral workshops to discuss nuclear force vulnerabilities and how best to manage different terrorist and insider threat scenarios has been proposed. It seems unlikely, however, that the Pakistanis would be willing to share much.[7] Destroying or retrieving Pakistani nuclear assets is another option that might prevent terrorists seizing them in a crisis. But the

United States would have extreme difficulty succeeding at either mission even assuming the Pakistani government invited U.S. troops into their territory.[8]

What else might help? If policymakers view the lack of specific intelligence on Pakistani nuclear terrorist plots against the United States as cold comfort and believe that such strikes are imminent—then, the answer is not much.[9] If, on the other hand, they believe conventional acts of terrorism and war are far more likely than acts of nuclear terrorism, then there is almost too much to do. In the later case, nuclear terrorism would not be a primary, stand alone peril, but, a lesserincluded threat—i.e., a danger that the Pakistani state could be expected to avert assuming it could mitigate the more probable threats of conventional terrorism and war.

What sort of Pakistan would that be? A country that was significantly more prosperous, educated, and far more secure against internal political strife and from external security threats than it currently is. How might one bring about such a state? The short answer is by doing more to prevent the worst. Nuclear use may not be the likeliest bad thing that might occur in Pakistan, but it is by far the nastiest. Certainly in the near- to mid-term, it is at least as likely as any act of nuclear terrorism. More important, it is more amenable to remediation.

This last point is the focus of this volume's first two chapters. Neil Joeck, now the U.S. National Intelligence Officer for South West Asia, and Feroz Hassan Khan, Pakistan's former director of Arms Control and Disarmament Affairs, examine just how easily conventional wars between India and Pakistan might be ignited and go nuclear.

The first observation both analysts make is that keeping the peace between India and Pakistan is now a serious issue for U.S. security officials. With 55,000 American troops in Afghanistan, Washington can ill-afford increased military tensions and nuclear rivalries between Islamabad and New Delhi that deflect or reduce Pakistan's own anti-terror operations along Afghanistan's southern border.

More worrisome is their second shared assessment: India and Pakistan have developed military doctrines that *increase* the prospects of nuclear use. Although India has pledged not to use nuclear weapons first, it has increased its readiness to launch shallow "Cold Start" conventional military strikes against Pakistan calibrated to deter Pakistani military or terrorist incursions. Meanwhile, Pakistani military planners insist that Pakistan will use nuclear weapons immediately if India attacks Pakistan's nuclear forces, conventional forces, and territory, or if it strangles Pakistan's economy.

Unfortunately, each of these countries' plans to deter war are too prone to fail. Precisely how does India intend to attack Pakistani territory either in a shallow or temporary fashion without tripping Pakistan's nuclear trip wires? The U.S. interventions, following terrorist acts that the Indian public has accused Pakistan's intelligence service of having backed, kept India from attacking Pakistan, but will such U.S. interventions work in the future? Indian military planners claim that they want to be able to punish Pakistan for any future perceived provocations well before any U.S. intervention has a chance to succeed.

Given India's interest in escalating its schedule of conventional military retribution, will Pakistan decide to intensify its own nuclear deployment efforts to persuade New Delhi that it is serious about its nuclear first use doctrine? How can Islamabad adjust its forward deployed nuclear forces to be credibly on the ready without also increasing the odds of unauthorized use or military miscalculation?

Then, there is the larger problem of nuclear rivalry. India claims the size and quality of its nuclear forces are driven by what China has; Pakistan, in turn, claims that the size and quality of its nuclear forces are driven by what India has. As one enlarges its forces, so must the other.

In an attempt to disrupt an action-reaction nuclear arms race while still ambling ahead, New Delhi recently persuaded the United States and other leading nuclear supplier states to allow India to expand its civilian nuclear and space launch sectors with imported foreign technologies and nuclear fuel. India's hope here is not to ramp up its domestic rocket and reactor production directly so much as to upgrade these programmes and free up and supplement its own domestic missile technology and nuclear fuel production efforts with peaceful foreign assistance.[10]

Although subtle, this approach has failed to calm tensions with Pakistan. Instead, Islamabad has used U.S. and foreign nuclear and space cooperation with India as an argument for enlarging its own nuclear arsenal. Thus, in 2007, Pakistan's National Command Authority warned that if the U.S.-India nuclear deal altered the nuclear balance, the command would have to reevaluate Pakistan's commitment to minimum deterrence and review its nuclear force requirements. Reports then leaked out that Islamabad had begun construction of a new plutonium production reactor and a new reprocessing plant. Shortly thereafter, Pakistan announced plans to expand its own civilian nuclear power sector roughly 20-fold by the year 2030 to 8.8 gigawatts generating capacity. The idea here is to expand Pakistan's ability to make nuclear electricity that would also afford it a larger nuclear weapons-making mobilization base it could use if India ramps up its own nuclear weapons-making efforts.[11]

This brings us to this volume's third chapter by Peter Tynan and John Stephenson of Dalberg Global Development Advisors. Just how economically sensible is expanding Pakistan's civilian nuclear sector over the next 2 decades? The short answer is not very. As Tynan and Stephenson explain in their analysis, "Even under Pakistan's most ambitious growth plans, nuclear energy will continue to contribute a marginal amount [3 to 6 percent] of electricity to meet the country's economic goals."[12] More important, building the number of large reactors that this level of expansion would require would be extremely difficult to achieve. Expansion of alternative energy sources, decentralized micro hydro, increased energy efficiency, coal, and natural gas, they conclude, would be far less risky.

In fact, they conclude that Pakistan could save considerable money over the next 2 decades and achieve its energy goals sooner by *not* building more nuclear power plants. The political salience of this point is magnified when paired with earlier analyses that Tynan and Stephenson did of India's planned nuclear power expansion. In India's case, Dalberg's conclusions were much the same: India could not meet its energy goals even under its most ambitious nuclear expansion plans, and there were a number of cheaper, quicker alternatives that make near- and mid-term investment in nuclear expansion a bad buy.[13] Bottom line: In both the Pakistani and Indian cases, expanding nuclear power only makes sense if one is willing to lose money or is eager to make many more bombs.

Judging from the state of its current finances, Pakistan can ill afford to do either. This much is clear from the economic analyses of Shahid Javed Burki and S. Akbar Zaidi presented in Chapters Four and Five. Pakistan, Burki writes, faces a "grim economic situation": "There is likely to be a sharp reduction in the rate of economic growth, an unprecedented increase in rate of inflation, a significant increase in the incidence of poverty, a widening in the already large regional income gap, and increases to unsustainable levels of the fiscal and balance of payments gaps."[14]

Moving the nation away from foreign charity funding toward an economic growth agenda will not be easy. Certainly, all unnecessary public spending, excessive military support, and consumer subsidies (e.g., for energy) must be cut. Pakistan, moreover, must assume a significant portion of the backend financing of its own planned growth. Investments in education and the agricultural sector must be increased substantially. Taxes will have to be increased without increasing the poverty rate or the already significant economic disparities between Pakistan's key regions.

None of this can come without political pains. To be specific, they will require political reforms that cannot simply be made top down from Islamabad, but will require a decentralization of powers to the localities. The good news is that some of this change may be pushed by modernizing trends, which both Burki and Zaidi note, are already under way. These include the urbanization of

Pakistan, the dramatic growth in electronic communications (e.g., cell phone use has increased 10-fold to roughly 50 percent of the population in the last 5 years, the number of private TV stations from one to more than 30), and the emerging domination of higher education by women (perhaps by a factor now as high as four-to-one) and their entry into Pakistan's work force.

In addition to these generally positive trends, there is increased investment in Pakistan and remittances from the oil-rich Persian Gulf, increasing trade with India (now Pakistan's seventh largest source of imports), and the prospect of a demographic dividend, which Craig Cohen details in Chapter Six. This demographic dividend, which will afford Pakistan a large labor supply relative to its young and old, Craig predicts, will continue to grow through the year 2050. This, he argues, has the potential to power significant economic growth "because the dependency burden is low," increasing savings and "allowing development of human capital."[15]

All of this should help stabilize Pakistan's economy and society. None of these trends, however, can possibly help if the government cannot reduce inflation (pegged at 28 percent in the first quarter of 2009), educate and feed its population, power its businesses and homes, and attend to its growing (and potentially violent) adolescent population. Achieving these objectives, in turn, requires political stability, domestic security, and increased domestic and foreign trade and investment.

It is unclear if this requisite stability will finally be achieved. What is clear, though, is that any successful attempt will only be possible if Pakistan and its friends focus less in the near term on direct forms of democratization and more on ethnic reconciliation and regional accommodation. Maya Chadda details how one might go about this in Chapter Seven. She makes a key recommendation that those assisting Pakistan—principally the United States—distinguish between violence that is driven by ethnic differences and that which is driven by Islamist terrorist organizations.

Professor Chadda argues that the United States should do more to help Pakistan integrate its ethnic groups while letting Pakistan and its military take the lead in fighting Islamic fundamentalism. What this requires, in turn, is an understanding of the key ethnic groups—the Punjabis, Sindhis, Pashtuns, Balochis, and others—and establishing metrics for safeguarding these groups' rights. Reforming Pakistan's federal model toward this end will not entail the promotion of direct, liberal democracy, but it will create the key building blocks necessary to create such a system. More important, it will give the key religious and ethnic groups the power and the interest needed to shape Pakistan's economic and social order and to keep them vested in Pakistan's future.

What, then, should the United States do? With regard to Pakistan reformulating its federal model, the United States might help to focus and condition economic assistance and freer access to U.S. markets and encourage Islamabad to foster greater equality among Pakistan's key regions and ethnic and religious groups. One suggestion that this book's authors discussed was giving each of Pakistan's provinces greater power to promote trade directly with India and focusing foreign investment to expand such commerce. The aim here is to moderate Indian-Pakistani relations by bolstering Pakistan's growing middle class. Pakistan, however, must take the first steps: If Islamabad does not want to reformulate its federal model to accommodate its various regions and ethnic and religious groups, Washington is in no position to help.

As for U.S. assistance to the Pakistani military, the key here is to do no harm. It is now fashionable in Washington to argue that U.S. policies toward India and Pakistan should be de-hyphenated. Yet, one sure fire way to increase Pakistani distrust of the United States and to increase its anxieties regarding India's military ambitions is for the United States to favour India's military modernization. If the United States wants to reduce the number of wars that could escalate into nuclear conflicts, it must make sure U.S. military aid to India and Pakistan does not prompt destabilizing military competitions. Accomplishing this, in turn, will require that the United States and other arms exporters provide these states with something other than mere quantitative equal treatment.

Consider missile defences. Because Pakistan has not yet fully renounced first use of nuclear weapons and India will always have conventional superiority over Islamabad, Pakistan would have cause to feel more insecure than it already does even if the United States or others gave equal levels of missile defence capabilities to both sides. In this case, India could diminish Pakistani nuclear missile threats and feel more confident about launching massive conventional military operations against Pakistan. Similarly, Pakistan would have far more to fear than to gain if the United States offers India and Pakistan equal amounts of advanced conventional capabilities since these might conceivably enable New Delhi to execute a humiliating "Cold Start" conventional strike against Pakistan's much smaller military or conceivably knock out Islamabad's limited nuclear forces without using Indian nuclear weapons. How the United States and others go about enhancing each of these states' offensive and defensive military capabilities, in short, matters at least as much as the actual quantity of goods transferred.[16]

While the United States should do all it can to discourage India from putting its conventional forces on alert against Pakistan, it also makes sense for Washington to make sure Pakistan's military and intelligence services stay focused against Islamist terrorist organizations operating in Pakistan. Here, it would be helpful to get India to reassure Pakistan that New Delhi is not supporting unrest in Balochistan and other areas in or bordering Pakistan. Yet another confidence-building measure that India should be encouraged to embrace is to invite the Pakistani military to all major Indian military exercises and to get Islamabad to reciprocate as part of a mutual military exchange. Finally, India and Pakistan should begin negotiations that would pull back forces identified to be offensive or threatening to agreed lines. No low, medium, and high-force zones could then be discussed. Here, North Atlantic Treaty Organization (NATO) conventional force reduction treaty expertise (including from Turkey) might be usefully tapped.

Making progress on any of these non-nuclear recommendations will help foster progress on the nuclear front. Here, the United States has a role to play in the implementation of the U.S.-India civilian nuclear cooperation agreement finalized in 2006. India may not be bound by the Nonproliferation Treaty (NPT), but the United States, Russia, China, and France—all NPT nuclear weapons states—are. Under Article I of the NPT, weapons state members cannot help any state acquire nuclear weapons directly or indirectly that did not already have them in 1967.

India no longer has any stockpiled uranium reserves. Shortly before the nuclear cooperative agreement was finalized with the United States, India was running its power reactors at a fraction of their capacity. That is one of the key reasons why India was so eager to get the United States to allow foreign uranium exports to India under the nuclear cooperative agreement. If India now imports a significant amount of nuclear fuel for its civilian power reactors, makes more nuclear weapons than it did before the deal, and does not increase its domestic production of uranium, it would necessarily be using the civilian imports of nuclear fuel to increase the amount of domestic uranium it could use to make bombs. This would implicate nuclear weapons states that might supply such fuel to India—e.g., Russia, France, China, and the United States—in violating Article I of the NPT. Under U.S. law, the Henry J. Hyde United States-India Peace Atomic Energy Cooperation Act of 2006 requires the U.S. executive to report annually on India's uranium consumption and supply to make sure that the United States is not implicated in any such violation. The idea behind the reporting requirement was to implement the U.S.-Indian deal in a manner that would threaten continued U.S. nuclear assistance to India's civilian programme should India use U.S. nuclear fuel imports to help it make more nuclear weapons per year than it was making prior to the deal. Promoting this approach with China, Russia, and France would clearly be useful: It could help restrain India's nuclear weapons materials production efforts and help the United States and the other NPT nuclear weapons states persuade Pakistan to do the same.

Ultimately, however, nuclear restraint by India and Pakistan is unsustainable without China doing more to restrain its nuclear weapons programmes and exports. President Barack Obama obliquely referred to this in his April 5, 2009, arms control address in Prague. After the United States and Russia agree to significant cuts in their nuclear weapons arsenals, he noted, "we will seek to include all nuclear weapons states in this endeavor."[17] For Pakistan's sake and that of South West Asia and the rest of the world, this endeavor should start as soon as possible.

One way to begin might be to encourage China to announce publicly what it claims privately to have already done—cease making additional fissionable materials for nuclear weapons. The United Kingdom (UK), France, Russia, and the United States already have made public statements to this effect and have made it clear that they have ceased this production as a matter of policy. If China were to follow, additional pressure might be placed on both India and Pakistan to do likewise. Certainly, it would be far preferable to attempt to balance the nuclear weapons efforts of Pakistan and India this way than by relying solely on the calibration of supplies to Pakistan and India of peaceful nuclear reactors, nuclear fuels, missile technology, and conventional military goods.

Of course, none of these steps will be easy. Each will take considerable time and effort. On the other hand, the reform agenda laid out here is far more tractable and concrete than anything flowing from concerns that Pakistani nuclear assets might fall into the wrong hands. Here, the specific options analysts propose are so extreme, they crowd out what's practicable. Rather than distract our policy leaders from taking the steps needed to reduce the threats of nuclear war, we would do well to view our worst terrorist nightmares for what they are: Subordinate threats that will be limited best if the risks of nuclear war are themselves reduced and contained.

Endnotes

1. See, e.g., Bruce Riedel, "Armageddon in Islamabad," *National Interest,* June/July, available from *www.nationalinterest. org/ Article.aspx?id=21644*; David Sanger, "Obama's Worst Pakistan Nightmare," *The New York Times,* January 8, 2009,

available from *www.nytimes.com/2009/01/11/magazine/11pakistan-t. html*; Evelyn N. Farkas, "Pakistan and Nuclear Proliferation," *The Boston Globe*, March 5, 2009; Paul McGeough, "West Warned on Nuclear Terrorist Threat from Pakistan," *The Age*, April 12, 2009, available from *www.theage.com.au/world/west-warned-on-nuclearterrorist-threat-from-pakistan-20090412-a40m.html*; The Editors, "Pakistan Nuclear Scenarios: U.S. Nuclear Solutions," May 5, 2009, *The New York Times*, available from *roomfordebate.blogs.nytimes. com/2009/05/05/pakistan-scenarios-us-solutions/*; Paul K. Kerr and Mary Beth Niktin, "Pakistan's Nuclear Weapons: Proliferation and Security Issues," Congressional Research Service Report for Congress, RL.34248, May 15, 2009, available from *www.fas.org/sgp/crs/nuke/RL34248.pdf*; and John R. Schmidt, "Will Pakistan Come Undone?" *Survival*, June-July 2009, p. 51.

2. See, e.g., Salman Masood, "Attack in Pakistani Garrison City Raises Anxiety About Safety of Nuclear Labs and Staff," *The New York Times*, July 4, 2009, available from *www.nytimes. com/2009/07/05/world/asia/05pstan.html?ref=world*. For a more detailed analysis of the connection between media emphasis on certain terrorist threats and terrorist operations, see Brian Jenkins, *Will Terrorists Go Nuclear?* New York: Prometheus, 2008. There also have been additional reports of other Taliban raids against Pakistani nuclear weapons-related sites whose veracity the Pakistani government has officially denied. See *Reuters* "Article Points to Risk of Seizure of Pakistani Nuclear Materials," August 12, 2009, available from *www.nytimes.com/2009/08/12/world/asia/12nuke.html*; and Global Security News Wire, "Pakistan Rejects Report of Attacks on Nuclear Sites," August 12, 2009, available from *www.globalsecuritynewswire.org/gsn/nw_20090812_6503.php*.

3. See, e.g., Helene Cooper, "South Asia's Deadly Dominos," *The New York Times*, December 6, 2008, available from *www. nytimes.com/2008/12/07/weekinreview/07cooper.html*.

4. See Abdul Mannan, "Preventing Nuclear Terrorism in Pakistan: Sabotage of a Spent Fuel Cask or a Commercial Irradiation Source in Transport"; Chaim Braun, "Security Issues Related to Pakistan's Future Nuclear Power Programme," and Thomas Donnelly, "Bad Options: Or How I Stopped Worrying and Learned to Live with Loose Nukes," all in Henry Sokolski, ed., *Pakistan's Nuclear Future: Worries Beyond War*, Carlisle, PA: U.S. Army War College Strategic Studies Institute, January 2008, pp. 221-368.

5. See Mark Thompson, "Does Pakistan's Taliban Surge Raise a Nuclear Threat?" *TIME*, April 24, 2009, available from *www.time.com/time/world/article/0,8599,1893685,00.html*; and David E. Sanger, "Strife in Pakistan Raises U.S. Doubts Over Nuclear Arms," *The New York Times*, May 4, 2009, available from *www.nytimes. com/2009/05/04/world/asia/ 04nuke.html?_r=1&pagewanted=print*.

6. See David E. Sanger and William J. Broad, "U.S. Secretly Aids Pakistan in Guarding Nuclear Arms," *The New York Times*, November 18, 2007, available from *query.nytimes.com/gst/fullpage.* html?res=9A07E5DC123AF93BA25752C1A9619C8B63&sec=&sp *on=*.

7. See, e.g., Rolf Mowatt-Larssen, "Nuclear Security in Pakistan: Reducing the Risks of Nuclear Terrorism," *Arms Control Today*, June/July 2009, available from *www.armscontrol.org/act/ 2009_07-08/Mowatt-Larssen*.

8. Although the United States has contingency plans to dispatch American troops to protect or remove any weapons at imminent risk of being seized by terrorists, the practicality of such a mission is extremely low. See R. Jeffery Smith and Joby Warrick, "Nuclear Aims by Pakistan, India Prompt U.S. Concern," *The Washington Post*, May 20, 2009, available from *www.washingtonpost.com/wp-dyn/content/article/2009/05/ 27/AR2009052703706_2.html*; and Thomas A. Donnelly, "Bad Options" in *Pakistan's Nuclear Future*, pp. 347-68, available from *www.npec-web.org/Books/20080116- PakistanNuclearFuture.pdf*.

9. See, e.g., James Jay Carafano, "Worst-Case Scenario: Dealing with WMD Must Be Part of Providing for Common Defence" *Heritage Foundation Special Report*, #60, Washington, DC: Heritage Foundation, June 29, 2009, available from *www.heritage. org/Research/HomelandSecurity/sr0060.cfm*.

10. See Zia Mian, A. H. Nayyar, R. Rajaraman, and M. V. Ramana, "Fissile Materials in South Asia and the Implications of the U.S.-India Nuclear Deal," in Sokolski, ed., *Pakistan's Nuclear Future*, pp. 167-218, available from *www.npec-web.org/Books/20080116-PakistanNuclearFuture.pdf*; and Richard Speier,"U.S. Satellite Space Launch Cooperation and India's Intercontinental Ballistic Missile Programme," in Henry D. Sokolski, ed., *Gauging U.S.-Indian Strategic Cooperation*, Carlisle, PA: U.S. Army War College Strategic Studies Institute, 2007, pp. 187-214, available from *www. npec-web.org/Books/20070300-NPEC-GaugingUS-IndiaStratCoop.pdf*.

11. Private interviews with senior Pakistani national security officials.

12. Peter Tynan and John Stephenson, "Is Nuclear Power Pakistan's Best Energy Investment? Assessing Pakistan's Electricity Situation," Henry Sokolski, ed., *Pakistan's Nuclear Future: Reining in the Risk*, Carlisle, PA: Strategic Studies institute, U.S. Army War College, Chap 3, December 2009, p. 99.

13. Peter Tynan and John Stephenson, "Will the U.S.-India Civil Nuclear Cooperation Initiative Light India?" in *Gauging U.S.-Indian Strategic Cooperation*, pp. 51-70, available from *www. npec-web.org/Books/20070300-NPEC-GaugingUS-IndiaStratCoop.pdf*.

14. Shavid Javed Burki, "Pakistan's Economy: Its Performance, Present Situation, and Prospects," Henry Sokolski, ed., *Pakistan's Nuclear Future: Reining in the Risk*, Carlisle, PA: Strategic Studies institute, U.S. Army War College, Chap 4, December 2009, p. 129.

15. Craig Cohen, "Pakistan 2020: The Policy Imperatives of Pakistani Demographics," Henry Sokolski, ed., *Pakistan's Nuclear Future: Reining in the Risk*, Carlisle, PA: Strategic Studies institute, U.S. Army War College, Chap 6, December 2009, p. 204, see fn 13.

16. On these points, see Peter Lavoy, "Islamabad's Nuclear Posture: Its Premises and Implementation," in *Pakistan's Nuclear Future*, pp. 129-66.

17. President Barack Obama, Arms Control Address, Prague, Czech Republic, April 5, 2009.

Chapter - 1

The Indo-Pakistani Nuclear Confrontation:

Lessons from the Past, Contingencies for the Future

—Neil Joeck*

Introduction

In 1998, India and Pakistan conducted a series of nuclear tests, making evident to the world and each other that they had a robust nuclear weapons capability. Despite the tests, the two countries fought a short war in 1999 and came close to fighting a second war in 2002. In both confrontations, the United States played an important role in helping to prevent escalation. The confrontations were followed by an extended diplomatic process called the Composite Dialogue that began in 2004 and served as a kind of umbrella for discussing the disagreements between the two sides. Given this history, it is likely that diplomatic dialogue and military confrontation will *both* play a role in resolving Indo-Pak conflict over the next several years. U.S. policies may also play a positive role in preventing crises from occurring and in mediating them when they do. This chapter reviews what happened in the two military confrontations and what lessons the two sides may have learned from them. It then assesses the implications of these conflicts for future crises, what scenarios may be envisioned for future conflict, and what steps the U.S. might take to reduce the prospects for nuclear use.

---

*The views expressed in this chapter are those of the author and do not represent the Lawrence Livermore National Security LLC or the U.S. Government.

In the end, India and Pakistan control their own future, but the United States can no longer afford to be a bystander in South Asia. Positive diplomatic developments over the past decade have resulted in the United States being engaged in South Asia on a permanent basis. It is unlikely—and probably not desirable for either India or Pakistan—that the United States would return to its historic pattern of paying attention to South Asia only in times of crisis. The effects of nuclear proliferation and international terrorism give Indo-Pakistani relations global consequence.

THE KARGIL WAR

What Happened?

In 1999, India and Pakistan fought a short war over disputed territory along the Line of Control (LOC) that separates their forces in Kashmir. It began in May when shepherds on the Indian side of the LOC encountered Pakistani infiltrators occupying land that had been vacated by Indian soldiers early in the winter. The commander of the army, V. P. Malik, was briefed on the incursion, but it initially appeared to be little more than normal artillery firing that characterizes the military confrontation along the LOC. Further reconnaissance, however, revealed a more widespread Pakistani occupation of key points around the town of Kargil. Although Islamabad claimed that the forces occupying the disputed ground were local freedom fighters, in fact Pakistan had deployed elements of the Northern Light Infantry into positions vacated by Indian troops, seizing a 200-kilometer stretch of territory. Once it was fully assessed, India saw that Pakistan's action significantly challenged India's control of the main highway through Kashmir and threatened to cut off resupply to India's forces based on the disputed Siachen Glacier. India escalated at the point of Pakistan's attack but, finding itself fighting up almost vertical heights, was unable to dislodge the invaders. When he was apprised by the Director General of Military Operations (DGMO) of events on the ground on May 15, General Malik advised that helicopters be brought into the battle, additional troops requested, and the Chiefs of Staff Committee (COSC) informed of the developments.[1] The Indian government then moved into crisis mode, established an ad hoc crisis committee, and escalated forcefully against Pakistan's positions.[2]

The ad hoc crisis team soon decided to take a step that upped the ante. India deployed air assets against Pakistan's entrenched positions, which India recognized could have "far reaching consequences" for Pakistan.[3] J. N. Dixit, a key member of the committee, saw the potential for serious military escalation: "... the use of the air force would change the nature of the military conflict . . . if India decided to deploy the air force in Kargil, India should be well prepared to anticipate the expansion of war beyond Jammu and Kashmir, and respond to expanded Pakistani offenses in other parts of India."[4] The implications of the decision were not lost on India. The use of air assets was an escalatory step, and Pakistan might, in turn, escalate still further. The war could expand beyond Jammu and Kashmir, which by definition would mean fighting across the international border.

Pakistan had started the war and showed no signs of giving up the fight on the battlefield; India was also prepared to escalate rather than back down. Were they prepared to do the same with their nuclear assets? The record is less open on this issue, but Malik notes that India had "one or two intelligence reports indicating that Pakistani Army personnel were noticed cleaning up artillery deployment areas and missile launch sites at the Tilla Ranges." Even though India had no specific information that Pakistan "was readying its nuclear arsenal... we considered it prudent to take some protective measures [and] some of our missile assets were dispersed and relocated."[5] On the other side of the conflict, Pakistan's Foreign Secretary, Shamshad Ahmed, stated on May 31 that Pakistan would not "hesitate to use any weapon in our arsenal to defend our territorial integrity."[6] Years after the war was over, an American official, Bruce Reidel, reported that on July 3, "more information developed about the escalating military situation in the area—disturbing evidence that the Pakistanis were preparing their nuclear arsenal for possible deployment."[7] The escalation to nuclear readiness appears to have been all too real.

As the war progressed, Pakistan's Prime Minister Nawaz Sharif grew increasingly nervous. He consulted with the United States and was told in no uncertain terms that his country had started the war, and it was his responsibility to end it. Strobe Talbott, then U.S. Deputy Secretary of State, later wrote that the United States "put the blame squarely on Pakistan for instigating the crisis, while urging India not to broaden the conflict."[8] After a hasty flight to Washington, DC, to consult directly with U.S. President Bill Clinton on July 4, Sharif returned to Islamabad and ordered the troops off the Kargil heights and back to their barracks. In his version of the war, Pervez Musharraf claimed that there had been no need for Sharif to recall the troops, that in fact they were holding up well and were prepared to continue fighting.[9]

Lessons and Consequences

Coming only a year after the reciprocal nuclear tests of May 1998, the Kargil War makes it clear that the acquisition of nuclear weapons did not prevent India and Pakistan from engaging in war. Indeed, nuclear weapons appeared to have little effect on Pakistan's planning. Only a small number of military leaders hatched the plan to seize the Kargil heights, and none of them apparently considered what role nuclear weapons would play. In a forthcoming volume on the Kargil conflict, the key planners appear not to have been dissuaded from their plan by the fact that India had demonstrated a fairly robust nuclear capability.[10]

In retrospect, the Kargil war appears to have contained a certain degree of mirror imaging, even though circumstances had changed dramatically with the acquisition of nuclear weapons. Pakistan's own response to India's similar seizure of contested territory along the Siachen Glacier in 1984 seems to have unduly influenced Pakistan's calculations. Pakistan had attempted to dislodge Indian troops from Siachen for several years, but finally decided that evicting the Indian troops would require a major offensive. Pakistan's military planners therefore assumed that India would draw the same conclusion regarding Kargil. Thus, Pakistan was surprised when India mounted a vigorous attack against Pakistan's positions and even escalated to the use of aircraft. This possible outcome was evidently never considered—nor was the corollary that escalation could continue to the nuclear level.

The role nuclear weapons may play, whether deployed or not, in deterring action or in sending a threat may not have been fully appreciated. President Musharraf argued that since "our nuclear capability was not yet operational... talk of preparing for nuclear strikes is preposterous."[11] This contradicts the threats implied by Shamshad Ahmed's comment noted above, but in any case seems to suggest that Musharraf believed that nuclear weapons only play a role when they are operationally deployed, without defining what deployment would entail. How were India's leaders supposed to know that Pakistan's weapons were not operationally deployed, and why would that knowledge lead them to conclude that their actions would not provoke a nuclear response? Are nuclear weapons only useful for intrawar deterrence? What lessons Pakistan drew remains to be fully explored.

On India's side, it is also not clear what role nuclear weapons played. India was not deterred from escalating at the point of attack and chose to mount a major offensive to regain the lost ground. Yet India's troops were under strict orders not to cross the Line of Control. That said, John Gill notes the "military and political leadership was careful to keep the option of cross-LOC operations open and used public statements by senior officials to highlight the latent threat of escalation."[12] Was the limit on crossing the LOC due to Pakistan's nuclear weapons? If so, why did the restraint not also apply to the use of aircraft? It is clear from Dixit's comment above that India knew that step could result in the possible expansion of the war. Yet they authorized the escalation.

On balance, it is difficult to reach firm conclusions about what lessons were learned about how deterrence worked at Kargil. Despite this, there is by now an assumption in Pakistan that Pakistan's nuclear capability forced India to fight a limited war, even though India was not deterred from escalating with respect to resources and was ready to fight across the international border if necessary. In addition, it is unclear how and whether limited war—typically defined in terms of limits on space, resources, time, and objectives—can remain limited in a nuclear environment. India's Chief of Army Staff (COAS) Malik points to the decision not to cross the LOC as a good example of how political control will ensure that wars in the nuclear age will not escalate. The decision to limit the war geographically but not in terms of resources, however, contradicts this optimistic assertion. Furthermore, India was prepared for escalation beyond the limits it initially intended to impose.

Finally, the duration of the war was determined by the Pakistani Prime Minister bending to U.S. pressure. From the perspective of the Pakistan military, however, the war could—and should—have continued. As President Musharraf writes, when asked by Prime Minister Sharif on July 3 as he was boarding his plane to Washington whether it would be necessary to accept a cease-fire and withdrawal, "My answer was the same: the military situation is favourable; the political decision has to be his own. . . It remains a mystery to me why he was in such a hurry."[13] Thus Kargil provides at best a mixed lesson in how war may stay limited under the nuclear cloud.

There was a lack of consensus among Indian and Pakistani observers about the outcome of the war as well as the influence of nuclear weapons. Pakistan felt that the military was acquitting itself well on the battlefield and had been sold out by politicians. Furthermore, Pakistan felt that its central objective had been achieved—Kashmir had been brought back to international attention. In contrast, India was convinced that Kargil was a victory for its own forces. India's troops had prevailed on the battlefield, India's political leaders had not been intimidated by Pakistan's nuclear weapons, and Pakistan had been portrayed to the international community as the aggressor.

A somewhat further worrisome outcome of the war is that Pakistan convinced itself that India was deterred from escalating because of Pakistan's nuclear weapons. In short, nuclear deterrence allowed Pakistan a certain freedom of action while it constrained India's response. This is a troubling conclusion if it inspires reckless actions in the future. It is doubly troubling in that the danger of nuclear escalation apparently did not affect the planners. If this betrays a belief that nuclear deterrence has an automatic character, it suggests that future conflicts may also be planned without due consideration of how the other side may utilize its nuclear and conventional capabilities.

A lesson that both sides seem to have drawn from Kargil is that although nuclear weapons do not prevent war, they do keep it controlled. Reason and hope suggest that this will always be the case, and the logic of nuclear deterrence supports such a conclusion. But people often act unreasonably and illogically, while wars have a way of turning out quite differently than initially planned. Therefore one cannot confidently cite the Kargil war as an example of how wars will be fought and whether nuclear weapons will remain in the background.

A final outcome of the 1999 war was the adoption by India of a nuclear doctrine that was introduced in draft form on August 17, 1999, and presumably was intended to inform Pakistan about how far it could and could not go in a conflict before it would face nuclear consequences.[14] It specified that India would develop a triad of delivery platforms. It stated that "credible, minimum nuclear deterrence" is a "dynamic concept related to the strategic environment, technological imperatives, and the needs of national security."[15] Thus it would have to change according to these factors, which would dictate the size, components, deployment, and employment of India's nuclear stockpile. The document specified command and control arrangements, research and development plans, and other elements of the overall decision structure. The key message it contained was that, ...any threat of use of nuclear weapons against India shall invoke measures to counter the threat and any nuclear attack on India and its forces shall result in punitive retaliation with nuclear weapons to inflict damage unacceptable to the aggressor.[16]

The message seemed to be that if Pakistan again threatened to use any weapon in its arsenal as it had during Kargil, India would respond likewise by readying its own weapons. If Pakistan used nuclear weapons, India appeared to be threatening the rough equivalent of the 1950s U.S. threat of massive retaliation.

Pakistan responded to India's nuclear doctrine with a challenge of its own. Three former senior foreign policy officials wrote a broad response to the new doctrine. Agha Shahi, Abdul Sattar, and Zulfiqar Ali Khan argued that India's new doctrine would threaten Pakistan's ability to respond.[17] In their view, India's declaration of a no first-use posture, if also adopted by Pakistan, would allow India to conduct a conventional first strike. Pakistan would therefore adopt a posture of flexible response and would use nuclear weapons first if necessary. The three authors specifically cited any attempt by India to occupy large parts of Pakistan's territory or to seize its communications junctions as causes for Pakistan to use nuclear weapons. In an interesting assertion, they claimed that nuclear deterrence had already worked. Once, in the mid-1980s when India decided against preventively striking Pakistan's nuclear installations: again in 1987 when an Indian military exercise threatened to boil over into cross-border war: and finally, in 1990 when Kashmir erupted in demonstrations following the kidnapping of the Kashmir Home Minister's daughter. This came as news to India and many analysts who did not see nuclear deterrence at work in any of these confrontations. The fact, though, that Pakistan considered nuclear deterrence to have prevented military action in those three instances underscored the lack of common understanding between the two sides about the role nuclear weapons played. It also begged the question why the planners of Kargil had paid so little heed to the role of nuclear weapons in their deliberations, while at the same time suggesting that Pakistan might take a number of provocative actions in the belief that nuclear deterrence prevented large-scale war.

Connected with India's nuclear doctrine was the recognition that India was not well-positioned conventionally to respond to the kind of war they had faced at Kargil. At an annual conference in New Delhi in January 2000 hosted by the prestigious Institute for Defence Studies and Analysis, General Malik presented the case.[18] He argued that India needed to find space between tolerating low intensity war of the kind Pakistan had fomented at Kargil and escalating to nuclear use. Defence Minister George Fernandes seconded this view, but no changes were made in force disposition or conventional planning. It would take another round of confrontation for India to address this challenge to its security.

THE TWIN PEAKS CRISIS

What Happened?

In a certain sense, the 2001-02 confrontation between India and Pakistan dates to September 11, 2001 (9/11) when al Qaeda attacked the United States, and Washington responded by sending troops into Afghanistan. For the first time since World War II, U.S. troops were on the ground fighting a war in South Asia. The cause of U.S. engagement was a global war on terror that Pakistan—after momentary reflection—had joined. Thus engaged, it would prove impossible for the United States to avoid getting caught in the middle of the Indo-Pakistani confrontation.

The actual Indo-Pakistani crisis began on December 13, 2001, when terrorists attacked India's parliament building, killing a number of guards but failing in their larger ambition of capturing and assassinating senior members of the Indian government. After examining the gunmen's dead bodies, India determined that the terrorists had been supported and probably directed in their actions from Pakistan. India responded by deploying upwards of half a million men along the LOC and the international border that divides the two nations. Almost immediately, however, India encountered enormous pressure from U.S. President George W. Bush and UK Prime Minister Tony Blair not to carry out its threat to retaliate for the attack on its Parliament.

Needing Pakistan's support for its operations inside Afghanistan, the United States was anxious to avoid a war in South Asia that would draw Pakistan's troops away from the Western border.[19] Washington placed numerous calls to New Delhi, urging Prime Minister Vajpayee to refrain from an attack. The United States argued that Pakistan would respond to U.S. pressure to stop infiltration across the LOC, so New Delhi should be patient. After a forceful personal intervention by Tony Blair and others, on January 12, 2002, Pakistan's President Pervez Musharraf went on nationwide television to denounce terrorism and call for a jihad against social ills.[20]

The speech closed a window of opportunity for India's decisionmakers. If they had a quick strike capability, it might have been used to counter Pakistan's apparent support for the terrorist attack against the Lok Sabha.[21] Instead, India was left to apply pressure as best it could under the strictures of its operating doctrine at the time. Called the Sundarji Doctrine for its author, General K. S. Sundarji, it deployed defensive, or holding, divisions near the border, with heavy strike corps kept in reserve for attack across the international border and deep into Pakistan. Getting this large force into position was a lumbering and time consuming process, ill-suited for a rapid response to a terrorist provocation. India was thus constrained from launching an attack against Pakistan in response to the attack on the Lok Sabha not only diplomatically and politically, but by the unwieldy nature of the build-up as well.

Despite India's conventional build-up, it appeared to Pakistan's leaders that nothing would happen because India was primarily focused on influencing the United States and the UK. In their view, the movement of forces was a substitute rather than a preparation for action. Even when terrorists attacked the Indian military camp at Kaluchak in May 2002 and ruthlessly murdered family members of the soldiers deployed along the LOC, India still held back. India's main demand throughout the confrontation was that cross-LOC infiltration must stop, which prompted a steady flow of diplomatic visits by high-level officials to Islamabad and New Delhi. This culminated in June, when U.S. Deputy Secretary of State Richard Armitage visited South Asia, stopping first in Islamabad and then in New Delhi. When he arrived in India, he declared that Pakistani President Musharraf had agreed to end such infiltration permanently. By summer's end, India declared that its objectives had been met, and the troops were returned to their barracks. The crisis had passed without any shots fired, but again with conflicting interpretations of the result.

Nuclear weapons were never at the forefront of the confrontation but were visible in the background. In January 2002, just as the two sides were close to completing their deployments, Khalid Kidwai, the head of Pakistan's Strategic Plans Division, which was in charge of Pakistan's nuclear weapons, granted an interview to two visiting Italian scholars. In his interview with Paulo Cotta-Ramusino and Maurizio Martellini, he sketched out four red lines that could prompt Pakistan to use nuclear weapons. They broadly repeated the two red lines Shahi, Sattar, and Khan had identified but added two more. Kidwai said that in addition to the territorial and communications (economic strangling) red lines, if India were to destroy a large part of Pakistan's land or air forces or destabilize Pakistan politically, Pakistan would be prepared to use its nuclear weapons.[22]

Then, following the Kaluchak incident, the nuclear threat became more palpable. On May 30, 2002, U.S. Ambassador Robert Blackwill ordered nonessential embassy staff and all dependents to leave the country. This was followed by an official State Department travel warning, implying that the possibility of war and of Pakistani use of a nuclear weapon against New Delhi was high enough that the United States could not justify endangering the lives of the embassy workers.[23] The UK issued a similar warning to its nationals in the area, and other Western governments duplicated the State Department announcement. India was outraged and privately accused the United States of capitulating to terrorism.

Despite its annoyance, the nuclear alarm may have had an impact. Although New Delhi had issued its draft doctrine following the Kargil conflict, a possible gap was made evident by the Twin Peaks confrontation. If India had invaded, as it was threatening to do, Indian troops might have found themselves inside Pakistan or Pakistan-controlled Kashmir. It was not clear from the doctrine, however, how India would respond to Pakistani nuclear use under those circumstances. This omission was addressed in January 2003 when the Prime Minister's Office issued a press release specifying that nuclear weapons would be used "in retaliation against a nuclear attack on Indian territory or on Indian forces anywhere" and in response to a biological or chemical attack on Indian forces.[24]

Lessons and Outcomes

From Pakistan's perspective, India had been bluffing through the whole process. Musharraf rejected India's assertions that Pakistan was connected to the terrorist attacks, and saw India's efforts as a failed attempt to drive a wedge between Pakistan and the United States. Pakistan had called India's bluff and demonstrated that any talk about fighting a limited war was hollow. In fact, as Islamabad saw things, war in 2002 did not need to be deterred because India never intended to fight a war. In the end, Pakistan stood firm, and India backed down.

From India's perspective, the U.S. "discovery" of the terrorist threat on 9/11 made Washington a Johnnycome- lately to the issue. Washington compounded the problem for India by aligning itself in the new global war on terror with Pakistan, who, in India's view, was the chief sponsor of terrorism. This misalliance handcuffed India after the Lok Sabha attack. India, the aggrieved party on December 13, 2001, and a victim of Pakistan's use of low-intensity conflict in Kashmir for a decade, was pressured by the United States to do nothing. The window of opportunity after December 13 closed on January 12, and Musharraf's speech was then used as a club to beat India. Because the United States wanted to fight its own war against terrorism in Afghanistan and needed Pakistan's help to do it, India was pressured to hold back. This may have made a virtue of necessity, since India at the time was saddled with the Sundarji Doctrine, but it was nonetheless galling to have to forgo a military response. Indeed, practicing restraint after the Kaluchak incident was very damaging to Indian civil-military relations, as the army was anxious to respond but was prevented from doing so for political reasons.

It was then even more disturbing to India to find the United States apparently knuckling under to veiled Pakistani nuclear threats. The decision to withdraw civilians from New Delhi demonstrated a craven lack of resolve that rewarded the perpetrator of terrorism while punishing its victim. Nuclear weapons seemed to have had a greater effect on the United States than on India itself. In sum, however, the 2002 confrontation, coupled with the problems identified in the Kargil conflict, revealed strategic weaknesses in India's defense policy and constraints on India's freedom of action that called for change and new thinking.

Just as 1999 caused new thinking in India about a nuclear doctrine, the 2002 confrontation made India take a new approach to conventional war. India had kept its response at Kargil limited geographically, but at great expense in terms of manpower. Its inability to mount a quick response to the terrorist attack in 2001 resulted in a costly and extensive build-up of conventional forces and also became a national embarrassment for the Indian army. Not only had what India declared to be Pakistani-supported terrorists attacked the symbol of India's democracy, they had murdered the dependents of soldiers preparing for a war that was never fought. Army post-mortems on the 2001-02 confrontation reached a number of conclusions. The Sundarji doctrine may have been appropriate in an earlier time for different needs, but it resulted in a slow motion and lumbering deployment of forces. It would have to change. In addition, Army analysts realized that even if the Sundarji doctrine were successfully implemented, it could very well cross key Pakistani red lines for the use of nuclear weapons. A new doctrine would have to account for Pakistani insecurities and avoid destabilizing intrawar deterrence. Finally, the new doctrine would also have to account for the intervention of third parties. A window for retaliation had been open from December 13, 2001, to January 12, 2002. The United States and the UK exploited this time to prevail on India's politicians and allow President Musharraf to evade the consequences of the terrorist actions. A new doctrine would have to enable India to strike on a very short time scale.[25]

The new doctrine was dubbed Cold Start and unveiled in April 2004. The idea was to restructure the Indian army so that it could address the defects made evident in 2002. With Cold Start, the ponderous holding divisions would be divided into eight or ten smaller integrated battle groups, each of which would be able to conduct shallow-penetration attacks across the border with Pakistan with relatively little leadtime required. This new doctrine would position the Indian army to conduct limited war against Pakistan, thus allowing New Delhi to retaliate against Pakistan swiftly before Islamabad could prepare militarily and before outsiders could intervene diplomatically, while also reducing the risk of escalation once the armies were engaged.

The contours of the Cold Start doctrine beg a number of questions regarding India and Pakistan's approach to limiting war. One of the more extreme interpretations of the objectives of the Cold Start doctrine would be the destruction of the Pakistan army.[26] This maximal position is almost certainly not endorsed by India's civilian leadership, nor by its entire military. Once introduced as a possible objective, however, Pakistan must treat it as at least a possible contingency that could become reality during conflict. Even if India explicitly rejected this objective, it brings up the problem of finding limits that both sides can accept and communicate. Suba Chandran makes the point that it is "essential to communicate to the other side the extent to which one would go in a limited war situation."[27]

In addition to communicating that the political objectives are limited, geographical limits will have to be identified. Borrowing from Thomas Schelling's discussion of tacit bargaining in a nuclear environment, India needs to ask whether new conspicuous stopping places can be mutually agreed once the LOC and international border are breached.[28] This may be difficult, as V. R. Raghavan argues "... there is no mutually agreed set of limitations between India and Pakistan on a future war—as there were none in past wars—neither side has control over the other's saliencies."[29] Pakistan has said that it would respond to a conventional Indian attack by escalating at the point of attack and expanding the war elsewhere at a point of its own choosing.[30] How will India and Pakistan agree on a new geographical limit once Cold Start has been implemented and either the LOC or the international border—obvious current limits, whose symbolism was reinforced in Kargil—have been breached? As noted earlier, one of Pakistan's red lines for nuclear use is territorial. If India attacks Pakistan and conquers a large part of its territory, Pakistan may respond with nuclear weapons. Implementing Cold Start without breaching this space threshold may be complicated once the bullets start flying. In addition to reaching tacit understandings about new geographical limits, they must also identify new limits on means during the induction of Cold Start. India breached the "no aircraft" understanding during Kargil. Communicating new limits while Cold Start is being implemented and Pakistan is escalating in response will be extremely difficult.

One of the obvious dangers as India plans how to conduct limited war is the prospect that Pakistan will be pushed to escalate. One of the goals of Cold Start is to avoid such an outcome, but it is difficult to predict outcomes once troops are engaged on the battlefield and new opportunities arise. The Shahi, Sattar, and Khan response to India's nuclear doctrine said that Pakistan would not use nuclear weapons tactically, and Pakistan has since indicated that it would use nuclear weapons in a relatively widespread attack. Given that any such use would compel India to respond in kind, leaving both countries devastated and rendering governance in Pakistan problematic at best, it is possible that Pakistan would reconsider how best to exploit its nuclear weapons during a war. Though highly fraught, the limited use of nuclear weapons might appear to be a better option for Pakistan if the alternative to nonuse was conventional defeat and the likely destruction of the state. The possible consequences of this new thinking and how conflict may again erupt in South Asia is discussed in the next section.

THINKING ABOUT FUTURE CRISES

Status of the Composite Dialogue

Following the 2001-02 confrontation, Pakistan and India reopened their political dialogue. At the annual meeting of the South Asian Association for Regional Cooperation (SAARC) in January 2004, Indian Prime Minister Vajpayee and Pakistani President Musharraf announced that they had agreed to resume the peace process that had been sidelined for several years. On February 18, 2004, they made the formal announcement that a bilateral "composite dialogue" would begin in the May-June 2004 time frame.[31] It is certainly too early to conclude that diplomacy has replaced conflict—the dialogue was suspended for several months in July 2006 after terrorists detonated as many as seven bombs on Mumbai trains—but diplomatic channels remain open, with the dialogue separated into eight different baskets. The baskets include Kashmir, peace and security, Siachen, Sir Creek, the Wullur Barrage, terrorism and drug trafficking, trade, and the promotion of friendly relations. These ministerial-level discussions have so far achieved varying degrees of success.

On Kashmir, a number of confidence building measures (CBMs) have been achieved and discussed, including the Srinagar-Muzaffarabad and Punch bus lines, crossing points on the LOC, and intra-Kashmir trade and truck services. The peace and security dialogue, held at the Foreign Secretary level (as are the Kashmir meetings), produced agreements on the prenotification of missile flight tests and nuclear accidents, a foreign secretary hotline and upgraded DGMO hotline, and reaffirmation of the ceasefire. It has not, however, been able to broach the issue of strategic restraint, leaving both sides unfettered as they increase their nuclear weapons stockpiles and expand strategic capabilities. The Siachen glacier dispute and Wullur Barrage remain contentious, but a joint survey of Sir Creek was agreed upon and may form the basis of a final settlement. Though no tangible results can be cited on drugs and terrorist issues, the two sides remain engaged and appear not merely to be casting aspersions on the other. Whether that spirit survives the deadly bombing of the Indian Embassy in Kabul on July 7, 2008, remains to be seen. Finally, the trade and friendly relations baskets remain subject to the political atmosphere and perhaps are notable for still proceeding as much as anything else. Pakistan remains concerned that India's tariff structure, especially regarding textiles, is too restrictive.

Though India and Pakistan are engaged in this structured dialogue, it is fragile and unlikely to weather any strong jolts. The Mumbai train bombings derailed it for a short period of time, but India came back to the table. Repeated attacks, however, could well force India's hand. The July 2008 terrorist attack on the Indian Embassy in Kabul, now determined by India to have been supported by Pakistan's Inter-Services Intelligence Division (ISID), is yet another example of the stress that is continually placed on the relationship and the efforts at diplomacy. So far, the desultory progress on the diplomatic front provides relatively little material gain to offset any sense on India's part of being played for a patsy. Conflict therefore cannot be ruled out, and a number of possible scenarios can be envisioned.

Conflict Scenarios: Triggering Events

A number of possible triggers for conflict are evident. The first and most obvious is another terrorist attack on an important economic or political symbol. The attack on India's embassy in Kabul was not sufficiently damaging to cause a crisis much less conflict. This can be a tricky issue for India. New Delhi wants to avoid intemperate and inaccurate remarks that would inflame relations with Pakistan at a time when India would like to see

Pakistan achieve political stability. It is not in India's interest to get in the way of Pakistan reaching a political accord that would stabilize its current government in Islamabad. At the same time, the experiences in 1999 and 2001- 02, as detailed above, make India want to avoid again appearing to be a passive and ineffectual victim of terror. Another attack on an important symbol or with significant loss of life may force New Delhi to act. It does not appear to be in Pakistan's interest to support any terrorist activity, but with Pakistan's military no longer running the country, there could be an increase in unauthorized activity by the army or the ISID. This might be justified internally as a means to assert the military's independence, to galvanize opposition to India's involvement in Afghanistan, or to force India's attention back to Kashmir. Furthermore, the terrorist organizations within Pakistan may well draw their own conclusions about what needs to be done regarding India. A violent action even two steps removed from ISID may be enough to compel India to go after the source rather than the immediate perpetrator of a terror attack.

A second possible triggering event would be the assassination of a key political leader. Political violence is regrettably common in South Asia, Benazir Bhutto's death only being the latest in a string that includes Rajiv Gandhi, Zia ul-Haq, Indira Gandhi, Liaquat Ali Khan, and Mohandas Gandhi. As with the Kabul bombing, there would have to be quite reliable evidence that Pakistan was somehow behind the killing for it to prompt an Indian response. Even in the absence of solid evidence, however, suspicions could lead to escalating tension, which itself could be a sufficient trigger. Another aspect of this factor would be the assassination of a lesser political leader such as one of the Kashmir politicians working with New Delhi, notably Omar Abdullah or Mehbooba Mufti. The likely resulting demonstrations and violence within Kashmir would inevitably increase tensions between India and Pakistan.

War could be instigated either in connection with or separate from an assassination of a prominent Kashmiri leader. Should militancy return to Kashmir, fanned by Pakistan or a spontaneous response to some real or imagined affront, it could take a more venomous form than previously seen. The demonstrations that followed the August 2008 decision by India to cede ground in Kashmir to Hindu penitents visiting the Amarnath shrine did not foment a new round of Indo-Pakistani conflict, but did make evident how tenuous relations are over this region. If Taliban or al Qaeda elements turned their energies to supporting Muslims in Kashmir, the outcome could be savage. Suicide bombing is now part of Pakistan's landscape—a few well-planned suicide bombings in Kashmir could easily trigger a dramatic Indian response across the LOC.

Another possible trigger for war may be India's Cold Start doctrine, whether it has been fully implemented or not. Pakistan is not inclined under current conditions to preempt as it has done in the past. In December 1971, when war was effectively already underway in East Pakistan in the form of Indian support for the Mukti Bahini guerrilla forces, Pakistan conducted preemptive air attacks against India's Western positions in the hope that India would engage in the West, where Pakistan held slightly better positions, and defer attacking in the East, where Pakistani forces were isolated and vulnerable. Pakistan did not preempt in 1987, however, even though India's Brass Tacks exercise began to look like preparation for an Indian attack against Pakistan.

Pakistan still sees itself as potentially vulnerable to an Indian armored attack, however, and although the Cold Start doctrine is intended to allay Pakistani fears that any of its red lines would be crossed in a conflict, it could well have the opposite effect. If Pakistan fears that it cannot rebuff Indian forces at all the points of attack envisioned in Cold Start, it may decide to take the initiative in a future crisis and launch an attack at a point of its own choosing rather than allow India to dictate the terms of a conflict.

War could also result indirectly from a coup by radical elements within Pakistan's army against the current moderate leadership. Again, this is an unlikely eventuality, but a new civilian government may well target the army and wish to punish it for the 9 years of army rule from 1999 to 2008. Former Prime Minister Nawaz Sharif has rhetorically asked why only civilians should be hanged, a clear reference to the military decision in 1979 to hang Zulfiqar Ali Bhutto and at the same time a threat to Pervez Musharraf for his role in the October 1999 coup and subsequent leadership of Pakistan. As Prime Minister from 1996 to 1999, Nawaz sought to neuter the various opposing centers of power in Pakistan—opposition parties, the Supreme Court, the

National Assembly, the media, the Army—and, if reelected, may try to do so again. If Chief of Army Staff Kayani were to accept civilian intervention as former COAS Jehangir Karamet did in Sharif's earlier term, more radical elements could attempt a putsch. The consequence would be enormous turbulence within Pakistan, possibly including the imposition of martial law, a step Musharraf was loath to take. Such a sequence of events could set the stage for rising tensions and accusations hurled at India, potentially setting the stage for a new round of conflict.

An unpredictable but possible trigger for conflict could be a nuclear accident. This would likely occur in connection with one or more other factors that had escalated tension, but if a nuclear accident occurred even during a minor confrontation, both sides might suddenly face the reciprocal fear of surprise attack. Even if it were during routine activities—an electrical fire at a nuclear weapon manufacturing site or a nuclear release at a reprocessing plant—the side responsible for the accident might try to cover it up. If that were successful, there might be no problem, but the probability of success would be low. Then the discovery of the cover-up would inject fear into the other side—if it was only an accident, why not admit it? If, instead of trying to cover up the accident, full disclosure was made, the other side might ask for more information to ensure that no harm was intended. It would be natural in such circumstances, however, to resist offering too much information, yet failure to be completely forthcoming would only exacerbate the situation, creating further tension.

If the accident occurred during the transfer of a weapon or a nuclear component to a safer storage area or to a site for mating with other components, tensions would escalate dramatically. Why was the transfer being made? How many other weapons were being transferred? How many were already transferred and ready for launch? Had intelligence that was previously considered solid now proven to be erroneous? Even if the exaggerated fears captured by such questions were *not* running through the minds of the decisionmakers in the opposite capital, it might well be assumed by the state that was moving the weapons that such thoughts *were* influencing the other side. And if they were, would it not make sense for the other side to ready its own weapons with as much haste as possible? How and whether such a cycle could be broken would depend on a host of psychological and political factors, all of which could be highly stressed by the unraveling events. India and Pakistan have addressed this issue by reaching an agreement regarding nuclear accidents (discussed subsequently in the CBM section). Full disclosure that an accident occurred does not necessarily solve the fears raised here, however, leaving this issue a potential source of tension and misunderstanding.

A last illustrative example of a possible trigger for war between India and Pakistan would be a substantial ethnic uprising in Pakistan. Pakistanis believe that India has in the past aided and abetted Balochi national aspirations. It is possible that a more coordinated uprising could take place in Balochistan, perhaps with support from India, anti-Punjabi Taliban elements, or al Qaeda terrorists. This would be unlikely without other contributing factors being present, such as an aggressive Islamabad government intervention in the Federally Administered Tribal Area (FATA). Support for a Balochi uprising might be used as a distraction to take forces away from the FATA or elsewhere. Pakistan might respond against India by another Kargil-like incursion or open support for terrorists in Kashmir. This congruence of events could force India's hand and provoke cross-LOC or international borders maneuvers to confront Pakistan.

Conflict Scenarios: What Would War Look Like?

War in the future might look much like war in the past. Pakistani support for proxy forces, primarily irregular militants operating outside government control, would most likely originate from Kashmir but conceivably could have a base in Bangladesh or Nepal. Pakistan's goal in supporting proxy forces would be to tie up the Indian Army as much as possible, bleeding and hectoring its forces to convince India that a diplomatic resolution to Kashmir on Pakistan's terms must be found. A parallel to this kind of conflict would be Indian support for the same kind of activity in Balochistan without the same longer-term objective of resolving Kashmir, but rather to make clear that two can play the same game with damaging consequences for Pakistan. India's goal would be to force Pakistan to deploy its forces away from other fronts, thus reducing Pakistan's ability to respond elsewhere on the IB or LOC.

These proxy efforts have been conducted in the past but without either side taking the war to the source of support across the border. Although Cold Start was developed in part to provide India with an ability to intervene in response to terrorist activities inside India, there are options short of Cold Start that could produce a different kind of war. Rather than invoking Cold Start as presently conceived, India could respond to Pakistani support of proxy war inside Kashmir by conducting a "punish and leave" strategy.[32] This might be an incursion by Indian special forces for no more than a 3-4 day period to allow the destruction of key training camps and supply routes. An alternative might be a "punish and stay" operation, more like the Chinese invasion of Vietnam in 1979 or the U.S. invasion of Cambodia in 1971. In both cases, the invasion force would be sent in for a fixed but significantly longer period of time, with the intent of disrupting the enemy's ability to continue resupply or staging. Both run the risk that Pakistan would expand the war elsewhere along the contested boundary. That, however, would force India to fight entirely defensively at points of Pakistan-initiated conflict, which could consequently reduce the dangers associated with Cold Start.

Conflict would look quite different if India invoked its Cold Start doctrine in response to a Pakistani provocation. Here there could be at least three broad variants: success on all the seven or eight fronts that Cold Start envisions; success on a few fronts and failure on the others; or failure on all the fronts. The latter outcome would create fewest problems from the point of view of escalation and nuclear use, but is also the least likely given India's superiority in conventional terms.[33] The second possibility might be a more likely outcome. Pakistan might choose to concentrate its forces at key defensive points to overcome India's thinned out forces that are called for by the Cold Start doctrine. Confronting Indian forces at a few critical choke points or in defence of vulnerable cities would be more important than stopping every one of the seven or eight points of attack. The result might be more like a stalemate, assuming that the successful

Indian offensives stopped after achieving the planned shallow penetration. Battlefield initiative might carry some commanders away, however, especially if they encountered light and only harassing resistance. Whether Pakistan would interpret a deeper penetration by a lighter force as crossing the territorial red line would depend on the dynamic circumstances at play elsewhere along the border. The most dangerous scenario would probably be the first, where Indian forces succeeded in surprising Pakistan and were able to penetrate along seven or eight fronts and then dig in and hold their positions. Seeing itself defeated along a broad swathe of territory would force Pakistan into making critical decisions about nuclear escalation.

Such a decision would also be forced on Pakistan's leaders in the event of an outright cross-border war such as India threatened in 2002. Whether the new Cold Start doctrine will be flexible enough to allow a massed invasion consistent with the earlier Sundarji Doctrine is not clear. But a powerful deep thrust into Pakistani territory at one or more points would likely overwhelm Pakistan and force it to counterattack elsewhere in a flanking manoevre. The dynamics of that kind of conflict would again be difficult to predict, but it is more likely that India would be able to prevail on the ground than Pakistan. In such a case, Pakistan would have to decide whether escalation to nuclear weapons would make any sense. How those weapons may be employed will be discussed in the next chapter.

Limited Nuclear Use Options

As noted earlier, India has declared that it will not use nuclear weapons first but reserves the right to retaliate against nuclear, chemical, or biological use against Indian forces anywhere. In the hypothesized scenarios depicted above, Pakistan is in most cases on the losing end of the conventional war and in contrast espouses a first use doctrine. It is therefore more likely that Pakistan would need to consider more carefully than India what nuclear steps it might have to take in certain dire circumstances. What might limited nuclear use look like?

Decisions would be made under duress, with troops backpedaling on the battlefield, and the international community using a combination of threats and rewards to induce Pakistan to show restraint. Under such circumstances, Pakistan almost certainly would first issue a threat to resort to nuclear weapons. It is popularly believed that Pakistan used public comments by Dr. A. Q. Khan in 1984 and 1987 to threaten India with nuclear weapons. His February 1984 interview with *Nawa-i-Waqt* came when India appeared to be considering a preventive strike against the Kahuta uranium enrichment facility, and his January 30, 1987 interview with the Indian journalist, Kuldip Nayar, occurred just at the close of the tense Brass Tacks face off.[34] These incidents may have been what Shahi, Sattar, and Khan were thinking of in saying that nuclear deterrence had worked in those years. The comments by Shamshad Ahmed during Kargil may have been intended to convey a similar warning, but all the pronouncements were somewhat veiled.

Under the conditions posited here, any threat from Islamabad would need to be far more official for it to have immediate effect. It would certainly have to be time-bound and specific—we will do X in place Y if Indian troops have not silenced their guns by time Z. To reinforce its seriousness, Pakistan would need to proceed with visible readiness steps, for example, moving truck convoys (both dummy and real) to potential assembly points, broadcasting the deployment of missiles armed with conventional and nuclear weapons at undisclosed launch pads (possibly communicating to third parties the coordinates of some of them to reinforce the point), and so on. Pakistan might also want to leave itself options to demonstrate resolve without starting a nuclear escalation.

A third step therefore might be to test a weapon to quicken the decisionmaking pace for India. That would require already having a weapon in place, which is highly unlikely, but could perhaps be done with a week's notice. Moving a weapon into position for such an eventuality would require substantial foresight by Pakistan, but is in the realm of the possible.

A fourth escalatory step—or third if a weapon had not been prepositioned in a test tunnel—would be to conduct a test in the atmosphere, perhaps on a missile fired toward the Arabian sea. Each of these steps would require a time lapse to allow India to see reason and stop its offensive – but at the same time, it may be difficult to stop the action on the battlefield in a timely manner. There could be a real problem of actions and threats overtaking the decision process in New Delhi. In any case, Pakistan would be forced to make a fateful decision whether to use one or more weapons against Indian targets. With the armies likely enmeshed and intermixed on the battlefield, dropping a weapon would require care to avoid also killing Pakistani soldiers. This could argue for using a weapon well behind Indian lines, but that could produce only marginal effect on the actual fighting. Pakistan might instead target a military base close to the front.

The next escalatory step would be a fairly largescale attack. It is possible to imagine steps short of such an attack as described above, but at some point Pakistan would likely see no reason not to attack with large numbers of weapons on a range of military and industrial—and potentially civilian—targets. There might be an effort made to avoid Muslims, but at such a dreadful point it would be quite difficult to practice much target discrimination. Any attack would be both destructive and suicidal since it would shatter any lingering caution on India's side, and a similar attack would almost certainly follow in response. Both sides would be left with unimaginable damage and a long and painful recovery. Depending on the extent of the damage, there could also be widespread but likely temporary (1 to 2 years) global consequences for food production, health, and the environment.[35]

WHAT COMES NEXT?

Current Confidence Building Measures

Over the years, India and Pakistan have agreed to a number of confidence building measures (CBMs) whose record of success is, to quote the Stimson Center, a prominent proponent and supporter of CBMs in South Asia, "spotty at best."[36] As noted earlier, they have also made some progress in the Composite Dialogue on additional CBMs. One of the longer lasting and more touted nuclear-related CBMs is the agreement reached between Prime Ministers Benazir Bhutto and Rajiv Gandhi not to attack one another's nuclear facilities. This arose out of the concern in the 1980s that India was planning to conduct an attack against Kahuta. To assuage Pakistani concerns, India proposed that the two sides exchange lists of their nuclear facilities and agree never to attack the listed sites. This CBM has held steady for almost 2 decades.

Another CBM that emerged from a crisis was the April 1991 Agreement on the Advance Notification of Military Exercises, Manoevres, and Troop Movements. Communication channels were available in January 1987 when the Indian Exercise Brass Tacks threatened to explode into war. Reciprocal misinterpretations of the other sides' movements—by Pakistan of the orientation of India's exercise, by India of Pakistan's responsive military positioning—created heightened tension that was resolved by diplomatic discussions.[37]

To avoid a repeat of that crisis, the two sides agreed on prenotification mechanisms, which have also so far been useful in maintaining military communications and reducing apprehensions.

As technical capabilities expanded, so too did CBMs. For example, in 1999 the two sides reached an agreement to prenotify each other of flight-testing of ballistic missiles. With the two sides having developed a fairly large suite of missiles, they have by now conducted an equally large number of tests. Given the close proximity of the countries and short flight times of missiles, this agreement has special value and has been used quite frequently. The existence of a CBM, however, does not guarantee that stability will follow. In April 1998, after Pakistan duly notified India and then conducted a test of its new Ghauri missile, India was sufficiently irritated that it went ahead with the decision to test nuclear weapons.[38]

Other CBMs are on the books but are not fully implemented. For example, one CBM created a hotline between the Directors General of Military Operations (DGMOs), but off-the-record reports indicate that the respective DGMOs can be reluctant to pick up the phone lest that act be interpreted as a sign of weakness in a time of tension—precisely when the hotline is supposed to come into play. There is a scheduled once-a-week call, but when the need is greatest, this CBM has been underutilized. There has been a hotline between the Prime Ministers as well, going back to 1999. But just as technology creates needs, it can eliminate needs, and we may see the day when the two Prime Ministers simply include one another in their "favourite five" on their cell phones.

Another prominent attempt at a CBM was the Lahore Declaration that highlighted Indian Prime Minister Vajpayee's historic bus trip to Pakistan in February 1999. This declaration sketched out a number of positive cooperative steps regarding nuclear stability, but as India's Foreign Minister later commented, the bus to Lahore got hijacked to Kargil.[39] The Kargil war broke out only a few months later, and it was soon evident that Pakistan's military leaders had been planning the intervention even as the political leaders were breaking bread—or *naan* as the case may be—together. That said, one of the elements of the Lahore memorandum was implemented 8 years later in February 2007 when the Agreement on Reducing the Risk of Accidents Relating to Nuclear Weapons was signed. This CBM specifically addressed the contingency hypothesized earlier where the other side might misinterpret a nuclear accident and trigger counter moves. A swift and complete explanation of any nuclear accident would certainly serve to dampen fears, but its implementation, if such an accident occurs, will require great transparency. Both sides will need to overcome their fears during a crisis, which may prove to be a test not just of this agreement, but of their political systems and national will.

A number of links between the two countries are regularly severed during crises, which creates opportunities for the two sides to show that they are improving relations when the severed links are finally reestablished. Such measures as foreign secretary meetings, air links, flag meetings between military commands, sports exchanges (especially of cricket teams), opening consulates, and ministerial-level talks are sometimes hailed as signs of improved relations and the restoration of confidence. That kind of progress may do little more, however, than set the stage for a new round of cuts to demonstrate anger when a new crisis begins to boil. In an odd way, such links may allow each side to blow off steam and send a message well short of conflict, thereby increasing crisis stability. To count them as CBMs, though, might cheapen the currency.

Once reestablished, however, the content of the senior level meetings can produce new opportunities, if not formal CBMs. The Foreign Secretaries met in June 2004 to discuss the sensitive issue of peace and security in Kashmir, and in September 2005 they discussed the overall peace process; the respective Commerce Secretaries met in August 2004 to discuss difficult trade issues; the Foreign Ministers and Prime Ministers meet regularly at SAARC, the United Nations (UN), and elsewhere for discussions. This may seem like cold comfort given the severity of their dispute, but some venue for discussion, if not resolution, is seen by both sides as positive, necessary, and for now, good enough.

Options for U.S. Support

Concerns about the safety and security of Pakistan's nuclear arsenal became particularly pronounced following the 9/11 al-Qaeda attacks. Although the al-Qaeda bases were across the border from Pakistan in Afghanistan, there was still a certain amount of concern about whether Pakistan's nuclear weapons were adequately guarded and contained safety mechanisms. President Musharraf wrote that after 9/11: "Every American official from the president on down who spoke to me or visited Pakistan raised the issue of the safety of our nuclear arsenal."[40] Pakistan had only established a robust command and control structure in January 2000, and at first it did not include a separate division for safety and security of the arsenal.[41] It did assign responsibility for the management and operation of its nuclear programme to the Strategic Plans Division, which served as the pivotal secretariat between the Strategic Forces Command in the field and the National Command Authority as the apex decision body. Pakistan and the United States were willing to consider areas for improvement and cooperation in this management structure.[42] This is a sensitive area that potentially impinges on the most secret aspect of Pakistan's defence structure, so any expansion of the reported cooperation will be limited and dictated by Pakistan as a sovereign nation.

The same sensitivities apply to India, but that nation has not faced the same scrutiny as Pakistan since it has not had the same relationship to the Taliban, al-Qaeda's ally and erstwhile host. When the issue of safety and security was broached in passing with a senior science and technology advisor to the Prime Minister, the topic was dismissed quickly with the comment that India has adequately taken care of that problem.[43] This is suggestive of the difficulty the United States may face in engaging India, but if old narratives can be avoided and a common approach considered, there is as much room for U.S.-India cooperation on security, if not yet safety, as there is with Pakistan.

Beyond the narrow management of the nuclear arsenal, the United States has vast experience, from mostly successful management of nuclear assets but some from grossly unsuccessful management practice.[44] It is therefore in a position to discuss best practices with Pakistan and India. Best practices can be interpreted in different ways, of course, but the strategic dialogues between Pakistan and the United States and India and the United States could include discussions of transportation safety, emergency search operations, personnel reliability standards, and alternatives for perimeter security. Although tricky from a protocol and NPT perspective, bringing the heads of India and Pakistan's nuclear management directorates to Omaha for meetings and discussions at the U.S. Strategic Command could be extremely instructive. Educational exchanges can also help, whether it is placing Pakistani and Indian officers in U.S. academic institutions or supporting American instructors in the staff colleges to teach a specific course or serve as a resource person for a specified duration. On the U.S. side, bringing Indian and Pakistani military instructors for a fixed term assignment with the National Defense University could be extremely interesting and create bonds that could serve U.S., Indian, and Pakistani foreign policy objectives.

India and Pakistan are not ready for any comprehensive cooperative threat reduction efforts. Indeed their view of what "cooperative," "threat," and "reduction" mean and imply may be at odds with views held in Washington. However, that need not prevent sharing experiences and approaches to improve understanding of the nuclear management challenges and perhaps improvement of the operations in the field. A variety of cooperative efforts are underway regarding technology transfer, including the megaports and container security initiatives, but they fall outside the compass of nuclear management. Weapon and materials accounting and control must be done by Pakistan and India on their own so long as they see their nuclear stockpiles as part of their defence programmes. Fissile material stockpile and production remain contentious topics at the Conference on Disarmament but remain high on the agenda for bilateral U.S. dialogues.

Some issues can productively be addressed in Track II fora, and, although there have been many over the years with mixed results, the effort is worth making. U.S. Government officials are willing to admit that certain issues (e.g., counterterror cooperation, nuclear stability, and regional conflict) can be difficult to discuss in official dialogues. A somewhat routine exchange of interagency-cleared talking points is necessary but can be productively supplemented with informal discussions among policy cognoscenti who are then able to identify problem areas and opportunities for policy development that might otherwise be missed.

Pakistan's greatest need at present in the area of conventional military hardware has to do with counterterrorism equipment and training. Some of the same equipment might usefully be transferred to India. A more interesting area has to do with U.S. considerations of transferring ballistic missile defense technology to India. Given that Pakistan is worse off in a defense-dominant world, it is unlikely that the transfer of defensive technology equally to both sides would solve Pakistan's concerns, even if the technology were being discussed. It cannot be overemphasized that defense does not serve Pakistan's interest, since, as the weaker power, Pakistan's threat to use nuclear weapons serves a legitimate security interest. In a defense-dominant world, Pakistan would again become vulnerable to Indian conventional superiority. As the stronger power, India continues to be interested in ballistic missile defense, and so far the United States has been open to the idea. India has already acquired some relevant technology from Israel and is considering Russian technology as well. Pakistan appears already to be responding to the possibility that India will acquire some kind of defensive system and can be expected to expand its offensive capability accordingly. In a sense therefore, providing defensive technology to India fuels the arms race, but the United States is not alone in that market.

Technology transfer in the area of nuclear management and operations is problematic. The United States so far has interpreted Article I of the Nuclear Nonproliferation Treaty in restrictive terms. The entire U.S.-India civil nuclear deal also brings up complicated issues of what is allowable, who are the legitimate end-users of the material, what restrictions must be enforced on internal transfer, etc.

CONCLUSION

The 1999 and 2001-02 confrontations could have been worse without U.S. intervention. That said, neither India nor Pakistan sees the U.S. involvement as an unadulterated good. Many in the Pakistani Army feel politicians, who were too quick to succumb to U.S. pressure, stabbed them in the back. Meanwhile, many in India feel that the U.S. fear of Pakistan's nuclear weapons and compulsion about al-Qaeda blinded it to the perfidious Pakistan regime, and therefore the United States unduly pressured India not to act in its own best interest. Thus another round of crisis or war between India and Pakistan will confront some of this lingering resentment. On balance however, the United States will be engaged and has constructed relations with both countries that at least until recently were as positive with both countries at the same time as anyone can recall. With a new civilian regime in power in Islamabad, though, U.S. influence cannot be assumed. The United States may be hard pressed to sustain the positive diplomatic atmospherics of the past 8 years, but must bend every effort to do so in order to preserve some ability to offer its own good offices in a future confrontation. This chapter has sketched out some dire scenarios for conflict in the future. Resolving the dispute will fall to India and Pakistan themselves, but they may see value in turning to the United States. Sound diplomacy and technical engagement can help make it politically tolerable within these two countries for the United States to play that role, if and when the time comes.

Endnotes

1. General V. P. Malik, *Kargil: From Surprise to Victory*, New Delhi, India: HarperCollins Publishers, 2006, p. 107; Pervez Musharraf, *In the Line of Fire: A Memoir*, New York: Free Press, 2006, pp. 87-98.
2. S. Paul Kapur, "Nuclear Proliferation, the Kargil Conflict, and South Asian Security," *Security Studies*, Vol. 13, No. 1, Autumn 2003, pp. 79-105; and S. Paul Kapur, *Dangerous Deterrent: Nuclear Weapons Proliferation and Conflict in South Asia*, Stanford, CA: Stanford University Press, 2007, esp. Chap. 6.
3. *From Surprise to Reckoning: The Kargil Review Committee Report*, New Delhi, India: Sage Publications, December 15,1999, p. 105.
4. J. N. Dixit, *India–Pakistan in War and Peace*, London, UK: Routledge, 2002, pp. 55-56.
5. Malik, pp. 259-260.
6. "Pakistan May Use Any Weapon," *The News*, Islamabad, May 31, 1999.
7. Bruce Reidel, "*American Diplomacy and the 1999 Kargil Summit at Blair House*," Center for the Advanced Study of India, Policy Paper Series, Philadelphia: University of Pennsylvania, 2002, p. 3.
8. Strobe Talbott, *Engaging India: Diplomacy, Democracy, and the Bomb*, Washington, DC: Brookings Institution, 2004, p. 158.
9. Musharraf, p. 96.
10. Peter R. Lavoy, ed., *Asymmetric Warfare in South Asia: The Causes and Consequences of the Kargil Conflict*, forthcoming, 2009
11. Musharraf, p. 98.
12. John Gill, "Military Operations in the Kargil Conflict," in Peter R. Lavoy, ed., *Asymmetric Warfare in South Asia*.
13. Musharraf, p. 97.
14. "Draft Report of National Security Advisory Board on Indian Nuclear Doctrine," Section 2.3, a, August 17, 1999, available from *meadev.nic.in/govt/indnucld.htm* and *www.pib.nic.in*.

15. *Ibid.*

16. *Ibid.*

17. Agha Shahi, Abdul Sattar, and Zulfiqar Ali Khan, "Securing Nuclear Peace," *The News*, Islamabad, October 5, 1999.

18. Malik, pp. 365-366.

19. Polly Nayak and Michael Krepon, *U.S. Crisis Management in South Asia's Twin Peaks Crisis*, Washington, DC: The Henry L. Stimson Center, 2006; and Alex Stolar, *To the Brink: Indian Decisionmaking and the 2001-2002 Standoff*, Washington, DC: The Henry L. Stimson Center, 2008.

20. The text of President Musharraf's speech is available from www.pakistantv.tv/millat/president/1020200475758AMword%20file.*pdf.*

21. The Lok Sabha, "House of the People," is the directly elected lower house of India's parliament.

22. Paulo Cotta-Ramusino and Maurizio Martellini, "Nuclear Safety, Nuclear Stability and Nuclear Strategy in Pakistan: A Concise Report of a Visit by Landau Network - Centro Volta," *Pugwash Online*, January 14, 2002, available from *www.pugwash. org/september11/pakistan-nuclear.htm.*

23. See Krepon and Nayak, pp. 34-35.

24. Press release from the Prime Minister's Office, available from *www.pib.nic.in.*

25. Walter C. Ladwig III, "A Cold Start for Hot Wars? The Indian Army's New Limited War Doctrine," *International Security*, Vol. 32, No. 3, Winter 2007-08, pp. 158-190.

26. See Subhash Kapila, "India's New 'Cold Start' War Doctrine Strategically Reviewed," South Asia Analysis Group Paper No. 991, New Delhi, India, May 4, 2004.

27. Suba Chandran, "Limited War with Pakistan: Will it Secure India's Interest?" ACDIS Occasional Paper, Urbana-Champaign: University of Illinois, August 2004, available from *www.acdis.uiuc.*edu/Research/OPs/Chandran/ChandranOP.pdf.

28. Thomas C. Schelling, *The Strategy of Conflict*, New York: Oxford University Press, 1960, pp. 67-74. 29. V. R. Raghavan, "Limited War and Nuclear Escalation in South Asia," *The Nonproliferation Review*, Fall/Winter 2001, p. 15.

29. Personal interviews with Pakistan Army officials in Islamabad, June 2003.

30. Historic Kashmir talks Agreed," BBC Online, January 6, 2004, available from *news.bbc.co.uk/2/hi/south_asia/3371107.stm.*
31. The term is from Michael Krepon
32. *The Military Balance 2008*, London, UK: International Institute for Strategic Studies, 2008, pp. 341-345, 349-351.
33. Interview with Dr. A. Q. Kahn, *Nawa-i-Waqt*, Pakistan, February 10, 1984; Kanti Bajpai, P. R. Chari, Pervaiz Iqbal Cheema, Stephen Cohen, and Sumit Ganguly, *Brasstacks and Beyond: Perception and Management of Crisis in South Asia*, New Delhi, India: Manohar, 1997, pp. 39-40, 106-107.
34. See Owen Toon *et al.*, "Consequences of Regional-scale Nuclear Conflicts," *Science*, Vol. 315, March 2, 2007, pp. 1224-1225.
35. The Stimson Center site provides a review of the concept of confidence building measures as well as a chronology previous Indo-Pakistani efforts; available from *www.stimson.org/southasia.*
36. See Bajpai *et al.*, chaps. 2 and 3.
37. Private conversation with former Indian NSA Brajesh Mishra, April 2008.
38. Jaswant Singh, *A Call to Honour: In Service of Emergent India*, New Delhi, India: Rupa and Co., 2006, pp. 200-229.
39. Musharraf, p. 291.
40. Private conversation in Islamabad, Pakistan, January 2000.
41. David Sanger and William Broad, "U.S. Secretly Aids Pakistan in Guarding Nuclear Arms," *The New York Times*, November 18, 2007, p. 1.
42. Private conversation in New Delhi, India, March 2008.
43. The transfer of live nuclear warheads on aircraft across the continental United States without the knowledge of anyone in the chain of command being only one egregious example.

Chapter - 2

Reducing the Risk of Nuclear War in South Asia

—Feroz Hassan Khan*

INTRODUCTION

The new international environment has altered the concept of national security. Threats to international peace and security now emanate not from strategic confrontation between the major powers, but from regional conflicts and tensions and the spread of violent extremism by nonstate actors, threatening nation-states from within and transcending state boundaries and international security. In recent years, the levels of security enjoyed by various states have become increasingly asymmetric—some enjoy absolute security, others none at all. This environment of security imbalance has forced weaker states to adopt a repertoire of strategies for survival and national security that includes alliances and strategic partnerships, supporting low-intensity conflicts, and engaging in limited wars and nuclear deterrence.

---

*Views expressed herein are solely the author's personal views and do not represent either the Pakistan government or the U.S. Department of Defence. The author is grateful to Lieutenant Commander Kelly Federal, U.S. Navy, MA National Security Affairs, from the Naval Postgraduate School, for contributing in substance, editing, and assisting with this chapter. The author also thanks Naeem Salik, former Director of SPD and visiting Scholar at SAIS, Johns Hopkins University, Washington, DC, for inputs and comments; and Ms. Rabia Akhtar, Ph.D. candidate at Quaid-e Azam University, Islamabad, Pakistan, for sending published research material from Islamabad.

South Asia has witnessed increased regional tensions, a rise in religious extremism, a growing arms race, crisis stand-offs, and even armed conflict in recent years. Nuclear tests did not bring an era of genuine stability between India and Pakistan, though military crises in the region did not escalate into full-fledged wars, underscoring the need for greater imagination to rein in the risks due to the fragility of relations between two nuclear neighbors in an increasingly complex set of circumstances.

Pakistan's primary and immediate threat now is from within. Its western borderlands are rapidly converting into a battleground where ungoverned tribal space in proximity to the porous and disputed border is degenerating into insurgency both to its east into Pakistan as well as to its west into Afghanistan. The al-Qaeda threat has now materialized into spreading insurgency in the tribal borderlands, which is taking a heavy toll on both Pakistan and Western forces in Afghanistan. The newly elected government in Pakistan has hit the ground running; but still mired in domestic politics, it has been unable to focus on the al-Qaeda and Taliban threat that is rapidly expanding its influence and targeting strategy. The most tragic aspect of this conundrum is the success of al-Qaeda in creating cracks of misunderstanding between Pakistan and the Western allies, while exacerbating tensions and mistrust between Pakistan's traditional adversaries, India and Afghanistan.[1] For example, Pakistan's security nightmare which perceives India-Afghanistan collusion in squeezing Pakistan is exacerbated, while the Indian and Afghan security establishments perceive Pakistani Intelligence malfeasance as perpetuating the Afghan imbroglio. Worse, the outcome of this confusion and blame generates real advantage for al-Qaeda and the Taliban. Any terrorist act that pits Kabul, New Delhi, and Islamabad against each other and intensifies existing tensions and crises also throws Washington off balance, allowing al-Qaeda and its sympathizers the time and space to recoup, reorganize, and reequip, and continue to survive.

The only silver lining in this unhealthy regional security picture is the slowly improving relationship between India and Pakistan, which has developed over the past 4 years. Though relations are tense and still fragile, there is a glimmer of hope in this overall crisis-ridden region. The dialogue process between India and Pakistan has been somewhat resilient in the face of significant setbacks and changing domestic, political, and international landscapes within each.

It is very improbable that a nuclear war between Pakistan and India would spontaneously occur. The history of the region and strategic nuclear weapons theories suggest that a nuclear exchange between India and Pakistan would result from an uninhibited escalation of a conventional war vice a spontaneous unleashing of nuclear arsenals. However, this region seems to be the one place in the world most likely to suffer nuclear warfare due to the seemingly undiminished national, religious, and ethnic animosities between these two countries. Furthermore, lack of transparency in nuclear programmes leaves room to doubt the security surrounding each country's nuclear arsenal and the safeguards preventing accidental launches. Therefore, discussions aimed at mitigating a catastrophic nuclear war in South Asia should focus mostly on the unilateral and bilateral anti-escalation measures Pakistan and India can take regarding existing issues. Additionally, each country's perception of its security is interwoven with the political, diplomatic, and strategic movements of the external powers that wield significant influence in the region. Coherent and consistent behavior that discourages conventional and nuclear escalation, although sometimes imperceptibly, is needed from the United States, China, and Russia. Without this, both Pakistan and India are unlikely to feel confident enough to reduce the aggressive posturing of their conventional forces over existing cross-border issues, leaving the escalation from conventional warfare to nuclear warfare a very real possibility.

This chapter focuses on the India-Pakistan nuclear rivalry, leaving Afghanistan-Pakistan issues and Pakistan internal threat dimensions for later discussion. It argues from the basic premise that nuclear war between India and Pakistan will most likely result from an escalating conventional war that must be prevented at all costs. Though the likelihood is remote, a nuclear exchange from an accident or an inadvertent release cannot be ruled out in a crisis. The stakes for a structured peace and security that reduce the risk of war that could turn nuclear are extremely high and linked to international security.

The chapter is organized into five sections. The first section gives a brief overview of crises and nuclear management in South Asia. The second section analyzes the likely causes of a nuclear exchange and possible scenarios. The third section evaluates the unilateral and bilateral steps that Pakistan and India can take with or without reciprocity. The fourth section examines the roles and influences of external powers in reducing risk and encouraging a peace and security structure in the region. Finally, the fifth section summarizes the key arguments and recommendations.

AN OVERVIEW OF CRISES AND NUCLEAR MANAGEMENT

During the Cold War, two sets of questions about security in the nuclear age were raised by some serious studies pertaining to the management of nuclear capabilities. The first set pertained to the performance of the command system in peace and war, and the second analyzed the dangers of inadvertence during a conventional war breaking out in Europe.[2] Since the end of the Cold War and the recession of strategic threats, the relevance of these dangers seems no longer important at the global level. Concerns about stability are now more applicable to individual regions where nuclear capability has emerged, especially in South Asia where a bipolar regional rivalry has changed the security dynamics, and violent nonstate actors have created the potential for triggering a war between two distrusting nuclear neighbors. It is essential to understand the differences between the Cold War era U.S.-Soviet nuclear tensions and the nuclear race underway in South Asia, as the latter is fraught with a long history of unsettled disputes, intense cognitive biases, and proximity.

During the gestation period of covert development of their nuclear weapons, India and Pakistan underwent a series of military crises. The occupation of Siachin Glacier (1984) and the Brass Tacks Exercise (1986-87) broke the uneasy spell of peace and tranquility that existed between the two neighbours since the Simla peace accord in 1972. During this period, both countries faced domestic political and separatist challenges, with each side accusing the other of abetting insurgencies.[3] By 1989-90, the third military crisis began with the Kashmir uprising and prompted U.S. presidential intervention for the first time. The 1990 crisis was the first of its kind where the nuclear factor played a role. Controversy still exists with conflicting claims of whether Pakistan conveyed veiled threats and engaged in nuclear signaling during the crisis.[4] These crises of the 1980s have since shaped the regional security dynamics, which were directly influenced by three intertwined dimensions. The first dimension was the end of the Cold War, which lowered the strategic significance of South Asia, thereby allowing the superpowers to disengage from the region. Second, the war in Afghanistan mutated into intraregional civil war after the Soviet departure. Third, the uprising in Kashmir evolved into a fullfledged insurgency in Jammu and Kashmir. In the center of all these dimensions was Pakistan. It was first to face the blowback of the Afghan war due to its decade-long involvement in Afghanistan and its vital security interests in both Kabul and Kashmir. In the changed geopolitical environment, Pakistan came under nuclear sanction (the Pressler Amendment) by the United States, which did not stop Pakistan's desire to match India's nuclear and missile developments. Nuclear sanctions, in particular, accelerated the ballistic missile race. As India flight-tested missiles, Pakistan, in a desperate search of suppliers to match India, sought a substitute for the F-16 aircraft, the delivery of which was stalled due to the nuclear sanctions. Pakistan looked east for its missile programme and eventually received both liquid and solid fuel technology transfers to enable a strong base to proceed independently. By the end of the century, India and Pakistan would possess a nuclear capacity sufficient to destroy the subcontinent.

In May and June 1999, Pakistan and India were engaged in a high intensity crisis at Kargil that was unprecedented in terms of its timing, nature, and intensity. The Pakistani opportunistic land occupation purportedly to improve its defences was no longer considered business as usual along the Line of Control (LOC). In the summer of 1984, India occupied the Siachin Glacier, left undemarcated in 1971, which triggered a series of crises along the relatively quiet northern part of the LOC.[5] The act triggered instability between nuclear-armed neighbours, unacceptable to a world that was now deeply concerned about the nuclear dimension. The crisis deepened as India vertically escalated the conflict using airpower and threatened horizontal escalation. Its diplomatic and information campaign succeeded both internationally and domestically in rallying support behind India. The opposite happened in Pakistan. The victory of having done something after the ignominy of the loss of Siachin and other posts was short lived. The Pakistani narrative and justification fell on deaf ears both domestically and abroad. Isolated and under severe sanctions, Pakistan's internal mechanisms collapsed into confusion and its army was forced to withdraw after the Pakistani Prime Minister dashed to Washington for help. The breakdown of civil-military relations and its consequences continue to affect Pakistan nearly a decade later.

The Kargil crisis of 1999 remains a highly controversial one for a number of reasons. One aspect was the nuclear dimension of the crisies. The U.S. intelligence community and policymakers believe that the Pakistani military made imminent preparations for possibly mating nuclear warheads with ballistic missiles. The Pakistani officials involved with such preparations deny any such actions or event.[6] Kargil is celebrated as a diplomatic success for the United States in crisis deescalation; however, this was a shocking blow to Pakistan and a clear manifestation of a U.S. tilt in India's favour, decidedly against Pakistan. With overt nuclear weapons capabilities, the paradigm of stability shifted. But new powers do not learn the shift instantly. Like the old, new nuclear powers take time to move up the learning curve. As Robert Jervis has argued in his work, nuclear revolution is a slow process.[7]

Although the crisis threatened prospects of peace and security, the foundations and potential for a structured peace were laid earlier in 1998-99. Under severe international sanctions, India and Pakistan were pushed into bilateral negotiations culminating in a summit from which the famous Lahore Declaration that encompassed the Lahore Memorandum of Understanding (MOU) was drawn in February 1999. The Lahore MOU recognized the nature of the changed strategic environment and laid down the basis of the potential peace, security, and confidence-building measures. The run up to the Lahore Declaration, however, was not without highly intensive U.S. engagements with both India and Pakistan. The team, led by Strobe Talbott, was composed of high-level teams of nonproliferation and arms control experts with extensive experience of Cold War negotiations. The U.S. experts, however, were unaware of the nuances of regional security compulsions, while the South Asian security managers and the civil and military bureaucracy were equally inexperienced in the logic, lingo, and implications of classic arms control that had evolved during the Cold War nuclear rivalry.

The Pakistani interaction with the United States (and dialogue with India) indicated a fast learning experience.[8] Substantive exchange of nonpapers with the U.S. teams led both sides to understand the obstacles and prospects of a minimum deterrence posture. Pakistan proposed the adoption of a Strategic Restraint Regime (SRR) for South Asia. The SRR was to consist of three interlocking elements: agreed reciprocal measures for nuclear and missile restraints to prevent deliberate or accidental use of nuclear weapons; establishment of a conventional arms balance as a confidence-building measure; and establishment of political mechanisms for resolving bilateral conflicts, especially the core disputes over Jammu and Kashmir.[9]

Of these three components, the two military elements were symbiotic and fundamental to Pakistan's security perspective and deterrent posture. The fundamental principle was a nexus between nuclear restraint and conventional force restraint. India dismissed the notion of conventional force restraints with Pakistan outright, indicating it would only discuss nuclear and missile restraint and doctrinal aspects. The U.S. experts were equally unenthusiastic. One interpretation was that linking conventional force restraints with nuclear restraints contained an implicit legitimization of upping the nuclear ante in the face of conventional threat. To the

Pakistanis, tying down the nuclear hand while freeing up the conventional hand was tantamount to legitimizing use of conventional force by India, and delegitimizing the use of nuclear weapons by Pakistan. What, then, was the logic of Pakistani nuclear deterrence that was achieved after 3 decades of opprobrium, sanctions, and military defeat in 1971—the original raison d'être for going nuclear?

The process of separated triangular strategic dialogue between each of the three—Pakistan, the United States, and India—created suspicions as each side was blind to the discussions of the other two. In Pakistan, suspicion especially grew for two reasons.

First, after 50 years of an alliance relationship with the United States, Pakistan was less inhibited in candor and trust. For India, this was probably new. However, U.S. sympathy and the public cozying up of Strobe Talbot and Jaswant Singh lent credence to onlookers that the United States was not interested in an equitable treatment of mutual restraint and potentially had a different agenda with India than with Pakistan. Second, the notion of dehyphenation was evident as the United States began to dismiss Pakistani security concerns; and, increasingly, U.S. negotiators began to mirror the perceptions and positions of their Indian counterparts.[10]

The strategic dialogue lost its seriousness, and soon it became a U.S.-India partnership dialogue rather than a U.S.-brokered chance of establishing a structure for regional stability. India was loath to accept any regional-based proposals as these would reduce India's status and elevate that of Pakistan.[11] Nevertheless, Pakistan took away many learning experiences. The dialogue process enabled Pakistan to set its priorities and align the key thinking on issues of doctrine, command and control, arms control, and nonproliferation concerns. In particular, the activities of A. Q. Khan crystallized the need for responsible oversight and restraint. There was a hiatus in the dialogue with the military government between 1999 and 2001. President Clinton's reluctant visit in March 2000 with the baggage of Kargil as the backdrop and a failed Agra Summit proved counterproductive in the end.

Encouraged by the success in Kargil and the U.S. response during negotiations, India announced its draft nuclear doctrine in August 1999, later made official in 2003. The draft nuclear doctrine, which announced the no-first-use policy, espoused a massive retaliation doctrine to include the use of nuclear weapons in the event of a major attack against India or Indian forces anywhere. If attacked with biological or chemical weapons, India would retaliate with nuclear weapons; and India supported this policy with the development of a triad of land, sea, and air nuclear weapons platforms. This was further enhanced by formal deployment of the Prithvi missile and subsequent development and deployment of the Agni series and other cruise missiles (Brahamos). On January 25, 2000, on the eve of India's constitutional birthday, Indian Defence Minister George Fernandes announced a doctrine of limited war under a nuclear umbrella. From a Pakistani perspective, every Indian pronouncement, India's doctrinal thinking, and its force goals and postures were directed at Pakistan specific interests and only indirectly referred to other unspecified threats (China).

In December 2001, just when U.S. forces were pounding at the Tora Bora hills to destroy the remnants of the Taliban and al-Qaeda, Pakistani armed forces were moving into the Federally Administered Tribal Area (FATA). Operation ENDURING FREEDOM had passed through a critical phase with Pakistan providing major logistics, intelligence, and operational space. Pakistani forces were required to be the anvil as U.S. forces were conducting operations across the region. This was the most crucial phase of the war against al-Qaeda for which the United States required major Pakistani military force deployment to block the porous border as best as they could. As military operations proceeded along the Afghanistan-Pakistan border, five alleged terrorists attacked the Indian parliament in New Delhi on December 13, 2001. This attack was the second of its kind within 2 months. The first attacks were on the State Assembly in Srinagar, Kashmir, on October 1, 2001. Enraged, India ordered complete mobilization of the Indian armed forces, and the Indian Prime Minister called for a decisive war against Pakistan. Since

1984, this was the fifth crisis and the largest and, at 10 months, the longest military standoff between the two rivals. This was also the first time that Pakistani armed forces were physically confronted on two battlefronts, particularly in the Spring of 2002 when U.S. forces conducted another follow-up military operation (Operation ANACONDA).[12] As brinksmanship and force deployment deepened on both sides, another terrorist incident occurred in May 2002, and war between the two neighbours seemed imminent. The consequence of the military standoff between India and Pakistan provided an opportunity for remnants of the Taliban and al Qaeda to escape into the porous borderlands with greater ease than would have been possible had Pakistan focused on a single front. The prospects of Pakistani force effectiveness in the tribal borderlands would have been greater at that time because tribal areas had up until then given no resistance to Pakistani force movement, allowing peaceful penetration into tribal areas. During the compound crises in 2002, India and Pakistan respectively signaled strategic unease through missile testing at two peak moments of their military standoffs. India tested its Agni-1 in January 2002, and Pakistan flight-tested three ballistic missiles in May 2002, prompting U.S. intervention to diffuse the crisis.[13] Given the propensity of crises in the region for the past decades, and with no prospects of conflict ending, there is not enough confidence that a miscalculation can be prevented in the future. The region refuses to acknowledge that limited or lowlevel conflict carries a threat of nuclear escalation.

POSSIBLE CAUSES OF A NUCLEAR EXCHANGE BETWEEN INDIA AND PAKISTAN

The legacy of suspicion created by violent events at partition still exists among many of Pakistan's and India's ruling elites.[14] Consequently, India and Pakistan have focused on internal balancing (i.e., modernizing their armed forces and eventually going nuclear) and external balancing (i.e., forging alliances or treaties of friendship with great powers).[15] This in turn contributed to the hardening of their respective stances on conflict resolution and the increasing frequency of cross-border crises. The nuclear capabilities of each only exacerbate the tensions inherent between the two countries, pushing each toward unilateral internal security-building measures. The double effect of the nuclear capability is that on the one hand it has contained crises and prevented major wars (deterrence optimism), but on the other hand, failed to prevent a series of military crises and dangerous confrontations (proliferation pessimism).[16] The mix of violent extremism and terrorism in the milieu has made regional security issues no longer an exclusive domain of any one state in the region.[17] Today, terrorist acts are not only affecting societies within the South Asian nations, but its effects ripple through the region and the world.

This section begins with the premise that surprise or unexpected nuclear exchange between the two countries is remote. This condition may change in the future for two reasons. One, change will happen if nuclear weapons are mated with delivery systems and deployed arsenals are routinely maintained, as was the case in Europe during the Cold War. Two, if strategic weapons asymmetry between India and Pakistan is broadened, it will increase India's first strike options in terms of capabilities, notwithstanding India's declared intentions of no first use in its official doctrine. This imbalance will occur in the future due to the introduction of destabilizing technologies and the freeing up of India's domestic fissile stock for military purposes, as and when the Hyde Act of 2007 is implemented.

Three major developments will erode the current balance in the future: Increasing capacities in advanced information, surveillance, and reconnaissance systems (Israeli-supplied Phalcon and Green Pine radars, for example); acquisitions of anti-ballistic missile (ABM) systems; and the steady militarization of outer space in which India has recently shown interest. Even if the possibility of a surprise strike against Pakistan may be remote and arguably meant for balance against China, these developments will force Pakistan into countervailing strategies. Pakistan's geophysical vulnerabilities to Indian aggression will increase, compared to China or any other country. This perceived invincibility against strategic arsenals would encourage India to wage limited wars with conventional forces. Since the 2002 military standoff and relative tranquility between India and Pakistan, the Indian military has experimented with new ideas of waging conventional war with Pakistan, as illustrated by the emerging military doctrine of Cold Start.

India's Cold Start military doctrine envisages creating multiple integrated battle groups that are self sufficient in limited offensive capacities—manoevere and firepower—forward deployed to garrisons close to Pakistan. One study suggests that the doctrine requires reorganizing offensive power of the three Indian Army strike corps into eight integrated battle groups, each roughly the size of a composite division, comprised of infantry, armor, and supporting artillery and other fire power units. This force would resemble the erstwhile Soviet Union's offensive manoeuvre groups, capable of advancing into Pakistan on different axes with the support of the air force and naval aviation.[18] The fundamental purpose of such a doctrine is to redress India's time-consuming mobilization of offensive mechanized forces, which loses surprise and allows Pakistan time to outpace India due to the short distances required for deployment. This was demonstrated in the 2002 crisis, and the Indian military was somewhat frustrated because of heavy-handed political control, diplomatic intervention, and loss of military opportunity to wage a short and limited, but intense, punitive war. Cold Start reflects several assumptions on the part of India. It dismisses Pakistan's nuclear capability, assumes accurate calculations of red lines, assumes it can control the degree of escalation, underestimates Pakistan's reciprocal conventional preparations and the subsequent retaliatory damage, assumes Indian and Pakistani governments will accept a fate accompli, and believes the reaction of outside powers (read United States) would be manageable and would help keep the conflict purely conventional and limited. These are all sizeable and significant assumptions; the failure of any opens the door to uncontrollable escalation to the nuclear level. The possible long-range outcomes for maintaining such a doctrine include an increasingly fortified India-Pakistan border, continued tension and pressure to maintain strategic weapons deployment, and a regional arms race. All three outcomes hinder the development of each country, but would be especially debilitating for Pakistan as it struggles to maintain two borders and a multitude of domestic crises.

Nuclear Force Deployment Scenarios

Should security dynamics unfold as described above, Pakistan will be forced to become a security state, far removed from the vision of a welfare state. In a heightened security environment with no peace prospects, there are four possible general scenarios in which Pakistan would be forced to consider deploying nuclear weapons, as outlined below:

1. Hot pursuit. India conducts punitive raids across the LOC or the international border. Imminent tactical preparations in India will force Pakistani conventional force reserves to mobilize.

2. Brass Tacks and composite crises 2002 revisited. Indian conventional force builds up for coercive deployment or decisive war (Brass Tacks or 2002 deployment), and nuclear forces are alerted and deployed.

3. East Pakistan revisited. India abets internal discords within Pakistan, inducing civil war, and seeks an opportunity to assail it as it did in 1971. Balochistan and parts of Sind and the North West Frontier Provinces, where domestic unrest and religious and tribal extremism are high, are good candidates for such a design.

4. Peacetime deployment of strategic weapons. India opts for formal deployment of nuclear forces, citing China or another strategic threat, and Pakistan follows suit.

The strategic picture profoundly changes should any conditions enumerated above manifest themselves. Lieutenant General Khalid Kidwai, in an interview with two Italian physicists, discussed hypothetical use scenarios and generally defines Pakistan's nuclear thresholds. Paolo Cotta-Ramusino and Maurizio Martellini quote Kidwai:

Nuclear weapons are aimed solely at India. In case that deterrence fails, they will be used if:

a. India attacks Pakistan and conquers a large part of its territory [space threshold],

b. India destroys a large part either of its land or air forces [military threshold],

c. India proceeds to the economic strangling of Pakistan [economic strangling], or

d. India pushes Pakistan into political destabilization or creates a large scale internal subversion in Pakistan [domestic destabilization].[19]

The four thresholds—geographic, military, economic, and domestic, as defined by Lieutenant General Khalid Kidwai—are factors that would determine the decision for deliberate use by a national command authority. These are not red lines, defined and understood by the adversary or other external parties. A clearly defined red line erodes nuclear deterrence and provides room for conventional force manoeuvre or destruction by firepower. The other possibility is inadvertent nuclear use. In this chapter, I use the Barry Posen model of inadvertent escalation and apply that model to the conditions applicable to South Asia.[20]

Nuclear Use Scenarios

In the absence of any structure of strategic restraint between nuclear-armed neighbours, the possibility of conventional wars breaking out is more likely. This then raises the question Barry Posen raised nearly 2 decades ago: the probability of inadvertent use. I argue that once the conventional war breaks out, the fog of war sets in and two major factors can create conditions for inadvertent use. First, during a conventional war, deceptions, countercontrol targeting, and communication breakdowns are routine consequences of warfighting. These elements contribute to the fog of war, which is further thickened by other conditions, as elucidated by Carl von Clausewitz in On War. Second, during peacetime, nuclear weapons safety is more important than effectiveness, especially if chances of war are small. But in war, the safety coefficient is of lesser significance than battle effectiveness. Again, this factor is not simply common sense, but critically important for deterrence stability. An unmated safe weapon will likely failsafe but is more vulnerable to preventive strikes. National command authorities cannot afford this risk and therefore must not only make weapons invulnerable but also capable of effective retaliation. It is the combined effect of these two factors that form the danger of inadvertence in the fog of war. As Martin Schram put it, "Danger of inadvertence is not guided by human planning but human frailty."[21] Following are possible scenarios that can cause inadvertence in the fog of war:[22]

Fog of War Scenario One: When strategic arsenals are deployed for war, deployed delivery vehicles capable of carrying both conventional and/or nuclear warheads are dispersed for protection and invulnerability. In addition, dummy warheads and real ones are mixed to deceive and keep the enemy guessing. The probability of misperceptions with the adversary increases, especially in South Asia. In the midst of war, any launch by such a strategic weapon (ballistic or cruise missile) will reach the target within 3 to 5 minutes. Depending on what warning and damage it does, any weapons fired from a strategic delivery vehicle will evoke unpredictable responses and the dimension of the battle will change.[23]

Fog of War Scenario Two: The second scenario could be derived from a communications break down in conjunction with a perceived rumour of decapitation or crippling of national leadership or the national command centers. Most modern wars commence with such a strike. Aircraft and ballistic or cruise missiles are ideal weapons to take out leadership in countercontrol strikes to decapitate nuclear forces, which are then either rendered incapacitated or incapable of effective retaliation. These forces, usually dispersed, camouflaged, and concealed, could then be neutralized by other means. In such an extreme case, for deployed nuclear forces to be effective, the "always" element of the command and control dilemma would become more expedient than the "never" element.[24] The last resort scenario would necessitate a "manual override" capability with nuclear weapon units.[25] This can only be undertaken in extremis, and it still does not necessarily imply that weapons units are independent or not under command or control of a formalized chain of command.

Fog of War Scenario Three: A conventional attack by aircraft destroys a nuclear weapon convoy or a fixed site on the ground, resulting in an explosion featuring a radioactive plume. In this case, it is unclear whether a nuclear weapon was used or the nuclear asset was blown up on the ground. Imagine a hypothetical scenario in which a Pakistani air force plane or ballistic missile were to hit an Indian nuclear weapon site or ballistic missile convoy. Will India construe this to be a first nuclear strike by Pakistan? Will India retaliate as enunciated in its strategic doctrine, or will India deliberate and evaluate what had happened before responding?

In all of the above scenarios, the best outcome would be that the respective national command authority does not jump the gun, assesses damage, and evaluates options. The worst case response, however, would be one made out of haste or impatience; war situations can cause irrational responses leading to an upward spiraling of panic within militaries and civil societies. The short flight times between countries suggest that this is a plausible scenario; therefore, the confusion and time-compressed reactions and responses in the heat of war should not be discounted. It is hard to predict the reaction and response of units in the field if some of their nuclear assets are destroyed or made ineffective by conventional attacks. In the ensuing chaos, would surviving units, if capable of operating manually, wait for authorization (enabling codes) and deliberation of the national command authority? Discipline, training, and Standing Operating Procedures (SOPs) would suggest they might; but as of yet, there is no precedent in history that sets a barometer to predict battlefield responses of militaries armed with both lethal conventional as well as nuclear weapons.

Pakistan's National Command Authority retains assertive control during peace and war. In a state of war, nuclear weapons will be mated with delivery systems; permissive action links to enable weapons will be established with a two- to three-man rule; and clearly articulated instructions about the authorization will be clearly issued to all commands.[26] However, it is unclear how command and control will cope with electronic jamming or other information warfare techniques that may preclude enabling weapon systems. Alternative command and communication channels are therefore always planned. In Pakistan, command and communication systems are wargamed each year to test the efficacy of the system. Even if redundancies fail, methods of establishing contact will be made through any means of transportation, including helicopters or ground transportation. Absence of communications will force local leaders to make use-it-or-lose-it decisions in case of severe attacks. However, should all other means fail, the last resort would necessitate pre-delegation to next-incommand or alternative commands as redundancy to assure retaliation, further enhancing deterrence.

UNILATERAL AND BILATERAL ANTI-ESCALATION STEPS FOR PAKISTAN AND INDIA

Unlike Pakistan, India is in a different position when it comes to reducing military tension between itself and Pakistan specifically and in South Asia generally. Its geographical size, central location, and military strength give India a hegemonic influence that it uncomfortably and inconsistently wields. In South Asia's turbulent history, India passed through its most dangerous decades relatively better than others, its smaller neighbours lacking adequate structure and strength to stem crises and wars. Regional security issues compounded also due to India's steadfast reluctance to accommodate its neighbours and to focus on a grand strategy of regional hegemony.[27]

India is still searching for the right strategy to deal with its neighbours, arguably impeding its own rise.[28] Two opposing schools of thought have emerged in the past 2 decades. The first school was based on engagement with its smaller neighbours on the basis of nonreciprocity, also referred to as the Gujral Doctrine.[29]

The second school of thought seeks a dominant posture and assertive policy towards neighbours, as enunciated in the Gandhi Doctrine.[30] India followed both tracks at various times, eventually favouring the hegemonic model. Had India pursued a broad approach of accommodation with its neighbours, it would not only facilitate better regional integration, but the prospects of fostering sustained peace and conflict resolution would be greater as well. A self-confident neighbourhood that has a stake in, rather than a fear of, India's rise is a harbinger for stronger structures of peace.

As identified above, India enjoys an edge in geophysical as well as qualitative and quantitative superiority over Pakistan. India can choose the time and place for an offensive, and it "is the conventional imbalance that could bring both sides to the nuclear brink."[31] Zawar Haider Abidi explains the Pakistani nuclear posture, which rejects the concept of no first-use primarily due to its perceived vulnerability to Indian conventional advantage.[32] A RAND Corporation study endorses the unlikelihood of a change in Pakistan's nuclear posture "without shifts in the conventional balance of forces, requiring CBMs [confidencebuilding measures] to demonstrate non-hostile intent" (e.g., halting training along the LOC in Kashmir or the prenotification of major military exercises).[33] As argued elsewhere in this chapter, the best pathway to assured nonuse of nuclear weapons is to undertake conventional arms control measures.

India and Pakistan go back a long way in negotiating treaties and elaborate CBMs.[34] Unfortunately, the record of implementation is rather unimpressive.[35] CBMs are no panacea for peace and security, but they are a useful foundation for potential structural arms control agreements. The basic reasons for the failure of CBMs is continuing distrust, aggressive force postures, forward deployment of military units, and continuing violence in the region. As one Indian author says:

"India has significant and identical CBMs with both China (stronger) and Pakistan (weaker) neighbours, the implementation of Sino-India and Indo-Pakistani CBMs have been different. With China, India has had positive experiences, with forces pulled back and tensions eased. India believes this is so because there is greater political will and common desire to normalize relations in the case of China, but not so in the case with Pakistan".[36]

The reasons go beyond the political will: India's and China's force deployments against each other are neither threatening in real time nor accompanied by active violence.

There is also a fundamental disagreement over the approach to peace and CBMs. India insists on transparency of doctrines as an important ingredient to tension reduction, particularly emphasizing a no first-use policy. Since Pakistan refuses to agree to such a step in the face of a superior conventional force, its diplomats concentrate on bilateral conventional and nuclear force reduction steps and India's offensive doctrines and force postures.[37] Subsequently, the process of agreement is extremely slow. Regardless, there are unilateral and bilateral measures the two countries can take to rein in the nuclear risks.

Unilateral Anti-Escalation Measures

Even though bilateral measures have the greater potential to reduce the likelihood of conventional escalation, there are steps each country can take without reciprocity, which could also mitigate escalation. On Pakistan's side, they can go beyond their ill-defined deterrence doctrine by specifically defining (and announcing) specific policies on key issues with appropriate parliamentary backing.

Strategic Weapons (warheads and missiles). Pakistan could make an official strategic doctrine that encompasses its concerns, doctrinal approach, and security obligations. Four main ingredients around which its doctrinal pronouncements could revolve are:

1. Minimum credible deterrence and eschewing of an arms race with India;

2. No first use of force—conventional or nuclear;

3. No transfer of nuclear technology to any state or non-state entity or provision of extended deterrence to any other country; and,

4. No use or threat of use of force against a non-nuclear state.

Strategic force postures. Pakistan can formally announce that unless the security situation dramatically deteriorates, its nuclear weapons will remain dealerted, its missiles and nuclear warheads will not be kept mated with delivery vehicles (aircraft or missiles), strategic weapons will remain operationally nondeployed, and Pakistan will provide notification of all missile tests. Islamabad should consider broadening its notification policy by including all neighbours of its tests, particularly Iran, Afghanistan, and China.

Conventional forces. Pakistan can formally announce it will not engage in a conventional arms race and will only maintain an acceptable ratio commensurate with its threats; and will not engage in dangerous hot pursuits, surgical strikes, or limited war with any neighbours across recognized borders or agreed lines of deployment (i.e., no more Kargils.)

Low-intensity conflicts. Pakistan should explicitly renounce the asymmetric strategies of using noncombatants in any shape or form as part of its security policy. It should explicitly announce that it will not allow its state territory or territory under its control to be used for training, organizing, preparing, and executing any form of cross-border violence (i.e., no more Operation GIBRALTAR or other forward policy as an extension of strategic depth). Pakistan should offer a joint regional terrorism cooperative center and open it to all neighbours and likeminded countries.

India, too, has some non reciprocity steps it can take to mitigate conventional escalation. The South Asian hegemon can unilaterally announce that it will neither cross borders or the LOC (i.e., no more Siachins), mobilize mechanized forces (i.e., no more Exercise Brass Tacks), or undertake coercive operations (i.e., no more Operation PARAKARMs) against South Asian Association for Regional Cooperation (SAARC) members, and that it will only maintain defensive formations within its border areas. This would preclude Brass Tacks-like developments and allow its smaller neighbours, Pakistan in particular, to focus their domestic military operations on counterinsurgency efforts. Furthermore and most importantly, India does have room to renounce its offensive military doctrines such as Cold Start and unlink future doctrines from the concept of limited war under a nuclear umbrella. Also, India's offensive military exercises can be reshaped not to portray Pakistan obtusely as the sole opposition force. Brazen Chariots, an exercise conducted in April 2008, is one such example that continues to harden Pakistan's belief that India's war preparations are Pakistan-specific. Lastly, India has the capacity to take the lead in coordinating joint military and naval exercises that support regional objectives, such as piracy reduction, expansion of search and rescue networks, and support of disaster relief contingencies. Such exercises not only expand the capabilities and skill sets of each country's militaries and actually improve the safety and security of the region, but they expand the breadth of relationships between rival countries, thereby lessening the chances of a conventional or escalating war.

Bilateral or Reciprocal Anti-escalation Measures

On the heels of the unilateral measures described above, previously hard-to-attain bilateral agreements will not be so daunting. And as far as reducing the risk of nuclear war on the subcontinent, bilateral and reciprocal measures will have exponentially greater success, making them essential ingredients to long-term nuclear stability. Since nuclear war will most likely be a result of conventional escalation, preventing military crises is the optimum goal of bilateral agreements and can be achieved through systematic steps.

First, India and Pakistan must agree to pull back forces that are identified as offensive and threatening to the other. This is not an untenable goal and, even if not entirely successful at first, can have a stabilizing effect. Merely getting together and pointing out what force postures are threatening will create an awareness of issues and attenuate the risk of inadvertently sending the wrong strategic message. After that could come the mutual creation of a "Low Force Zone" in which force deployments will be mutually negotiated and a "No Offensive Forces Zone" as appropriate.

The next series of bilateral steps would focus on the nuclear weapons themselves. However, such achievements are unlikely without outside support for such moves, particularly from the United States and China, and will therefore be discussed in the next section.

THE ROLES AND INFLUENCES OF EXTERNAL POWERS IN REDUCING RISK

Unfortunately, the influences of the United States, China, and Russia have not historically been consistently beneficial to the stability of South Asia. The superpowers have notoriously applied military and diplomatic pressure upon Pakistan and India when and where it seemed to best oppose the converse efforts of the adversaries, regardless of the effects it had on the civilians and governments that bore the brunt. Aid and technology was granted and denied to South Asia based not on the long-term regional stability implications, but on the respective central government's perception of its own periphery threats and its ability to provide such support. As the tides of support ebbed and flowed, South Asian countries redirected their solicitation as needed.

The U.S. military and economic support was particularly critical to Pakistan's survival, but the United States lent support to India when it was in its own interest, as during the 1962 war with China. In addition, the United States has played a significant role in deescalating Indo- Pak crises a number of times. Invariably, the regional countries looked towards other partners, namely Russia and China, when the expected U.S. support did not measure up or materialize.

The United States still exhibits the same pattern of behaviour. In the decade following the end of Cold War, it abandoned Pakistan in favour of connecting with a rising India, only to return to Pakistan after September 11, 2001 (9/11).[38] Seven years later, the United States is in an unprecedented position of influence in New Delhi, Kabul, and Islamabad, each an important partner in its own right and significance. However, the mutual suspicions in the ongoing regional rivalry compound regional and global security prospects and, worse, help enemies such as al-Qaeda.

A contention of this chapter is that the prevention of war between India and Pakistan is intrinsic to war against al-Qaeda—a hostile Indo-Pak relationship, particularly if it escalates toward force mobilizations against each other, hampers the U.S.-led war on terror. The U.S. policy has been to prevent nuclear weapon acquisition by war-prone states, and, if that fails, to prevent wars between nuclear-armed states. However, the India-Pakistan rivalry has direct impact on the most crucial security issue in contemporary times and all efforts to prevent nuclearization have failed, mandating a change of tack for the states wielding influence in South Asia. The United States, China, and Russia should proactively engage in three areas: (1) conflict resolution among all states; (2) strategic weapon threat reduction between India and Pakistan; and (3) conventional arms control between India and Pakistan.[39]

Conflict Resolution

The United States will need to expend a huge amount of time, energy, and money to bring Pakistan, Afghanistan, and India into a mode of conflict resolution, which is hampered by anti-U.S. sentiment in all these countries. But it is time to override objections and find a way to convince India that concessions made in the name of conflict resolution neither reduce India's status nor undermine its ambitions. Chinese involvement can serve to assuage fears of U.S. imperialism or overreaching while also providing a hegemonic stability upon which secure regimes can be constructed. The new U.S. administration should soon consider a Madrid-like process for South Asia.

Strategic Weapons Threat Reduction

It will likely be futile for the United States to work on lowering strategic force goals, as past experience has indicated resistance from both India and Pakistan. It would be more pragmatic to help India and Pakistan formalize nondeployment plans for their strategic weapons, dissuade the introduction of nuclear and non-nuclear destabilizing technologies, and assist in best practices for their nuclear regimes. Specifically, international actors should encourage Pakistan and India along the following four areas:

1. *Nuclear Risk Reduction Centers (NRRCs).* The basic purpose of NRRCs in each capital will be to have a focal point to prevent an impending crisis from escalating. Outside countries can join in to help establish such centers. The United States can play a vital role in encouraging nuclear and political confidence building measures.
2. *Personnel Reliability Programme (PRP).* Sharing experience on organization best practices such as PRPs and procedures to manage sensitive technologies will help respective national command authorities adopt most stringent practices of safety, security, and reliability. As mentioned, training and selection of personnel to withstand psychological pressures in the fog of war will be of the utmost importance in the crisis-prone region.
3. *Accident Avoidance.* The United States, China, and Russia all have a role to play in the realm of accident avoidance since they provided much of the original technology in use in South Asia. They could also share and possibly train a core of people on accident avoidance techniques and reduction of technological errors, such as electromagnetic radiation and computer fallibility.
4. *Physical Protection Technology.* The use of some generic physical protection and material accounting practices such as sophisticated vaults and access doors, portal command equipment should be mutually agreeable.[40] Again, there is sensitivity in both countries to such intrusion, so this cooperation must remain within the bounds of general training and allow countries to develop their own technology if desired.

Conventional Force Restraints

There are three principal reasons for a U.S. role in conventional force restraints in South Asia. First, between 1999 and 2001 the United States was the main supplier of sophisticated technologies and state-of-the-art platforms to the region. It must understand how this affects regional strategic instability, and why the need for conventional agreement is necessary. Second, the U.S. prime concerns are on the Pakistan–Afghanistan border. The United States expects and desires Pakistan armed forces to focus their military power on this all-important front—an unlikely occurrence absent a force restraint agreement with India. Third, the United States needs to examine not just the physical postures and build up of conventional forces but emerging military doctrines (Cold Start and low intensity conflicts/proxy wars) under the nuclear umbrella. These strategies undermine U.S. objectives of war against al-Qaeda.

The United States should encourage the development of overarching principles of identification, mechanism, and non-aggression agreements coupled with strategic weapon restraints. It would make sense to proceed gradually and simultaneously on parallel tracks towards conventional force restraint. Four stages of a conventional arms agreement can be brokered:

1. Identify offensive and defensive forces and requirements for other security forces.
2. Agree on designation of a determined "Low Force Zone." Any increase in strength equipment or structure is voluntarily made known to each other under a CBM.
3. Engage in restructuring and relocation of offensive conventional forces so as to build confidence and trust as other peace objectives are achieved.
4. India and Pakistan must engage in proportional force reduction efforts similar to the pattern of Mutually Balanced Force Reductions (MBFR).

In addition to the objectives outlined above, Chinese actions carry some added weight. Whether or not China builds up its nuclear capability based on South Asian security concerns or outside influences, it upsets whatever balance India might feel it has regarding the Asian power. The U.S. reliable replaceable warhead (RRW) programme exemplifies this. Although China may feel its 200 nuclear warheads is an adequate balance to the 10,000 U.S. warheads, the RRW threatens that balance and could cause escalatory ripples in South Asia via China.[41] Although

Chinese-Indian interaction has become increasingly positive and more frequent as of late, China's internal force posturing, especially in the nuclear realm will invariably create waves in India and, in turn, Pakistan. Support for Pakistan has become less overt under the scrutiny of U.S. military involvement in the area, but China also needs to keep in mind the indirect effect China has on the subcontinent when it starts altering the status quo of its forces.

KEY ARGUMENTS AND RECOMMENDATIONS

A nuclear-armed subcontinent is a reality. The best way to achieve strategic stability in the region is by establishing a structural peace and security framework for conventional war avoidance and formalizing the nondeployed status of nuclear weapons. Recent history has shown that reliance on the nuclear umbrella sheltering South Asia seems to have provided militaries on both sides of the border more strategic room with respect to perpetuating low intensity warfare and escalating conventional warfighting doctrines. Additionally, this chapter has argued that the most probable cause of a nuclear exchange on the subcontinent will be a result of conventional war escalation—either through accident in the fog of war or due to established protocols—and less due to accidental launches. Preventing a nuclear exchange in South Asia is, therefore, less dependent on strategic weapons safeguards, although they remain a key to strategic stability, rather more dependent on the prevention of conventional war escalation. Conventional, and therefore nuclear, stability can start through unilateral steps taken by Pakistan as well as India. More importantly, India as the primary regional power has significant responsibilities in preventing nuclear war and initiating anti-escalation measures. Where real stability will be achieved, though, is through bilateral and multilateral strategic actions improving the safeguards and reducing the apparent threats to opponents, superimposed by coherent superpower policies and involvement.

Because of India's primacy in South Asia, it must take the lead initiating stability-inducing policies and doctrines, particularly due to its relative military strength. Its behaviour has not been consistent over time, vacillating between accommodating (Gujral doctrine) and confronting (Indira Gandhi doctrine) in its dealings with other South Asian neighbours. India has leaned towards the latter as new international trends like Asian power balance and globalization, for instance, favoured India leaving little incentive for the former model.[42] Shifts in the international system—global terrorism, globalization, and informational and economic interdependence—will make traditional security issues less relevant. Regional security issues in South Asia are now qualitatively different and interrelated such as energy, water, food, poverty, terrorism, and rising religious extremism. India must take the lead.

A structured peace and security regime between India and Pakistan is now a geo-political compulsion. A cooperative relationship between India and Pakistan is directly related to peace and stability in Afghanistan. Unless India and Pakistan stabilize their relationship and change the culture from confrontation and exploitation to cooperation and collective gain, success in the global war against al-Qaeda will remain elusive.

The United States, in concert with major powers, can turn this grim and seemingly intractable security situation into a unique opportunity of security paradigm change from suspicion and rivalry to one of conflict resolution and stability. The stakes of preventing war and crisis between India and Pakistan (and Pakistan and Afghanistan) is now an extremely important ingredient of the global war on terror and is not just simply a matter of moving toward a peace between two nuclear-armed countries.

Nuclear neighbours with a long history of unsettled disputes, cognitive biases, crises, and wars require years of crisis-free confidence and trust building to mature into détente, aided by a supportive international community. Conditions for instabilities will continue so long as the dangerous trend of seeking space for low level conflicts continues, and the feasibility to wage limited conventional war under the nuclear threshold is not taken off the table. Nevertheless, as has been shown in this chapter, there are unilateral and bilateral steps India and Pakistan can take to rein in the risk of nuclear war on the subcontinent.

Endnotes

1. On July 7, 2008, a suicide car targeted the Indian embassy in Kabul, killing many including the Indian defence attaché. This terrorist incident has triggered angry responses from people in New Delhi, India, and Kabul, Afghanistan, who, not surprisingly, are pointing fingers at the Pakistani Inter-Service Intelligence. Relations are tense within the region. (At the time of this writing, September 13, 2008, in yet another terrorist incident in New Delhi, India, five blasts killed over 20 and injured dozens.)
2. See, for example, Ashton B. Carter, John D. Steinbruner, and Charles A. Zraket, eds., Managing Nuclear Operations, Washington, DC: Brookings Institution Press, 1987; Desmond Ball, "Can a Nuclear War Be Won?" Adelphi Papers 169, London, UK: The International Institute for Strategic Studies (IISS), 1981; and Barry R. Posen, Inadvertent Escalation: Conventional War and Nuclear Risks, Ithaca, NY: Cornell University Press, 1991.
3. India was involved in Sikh, Tamil, and Naxalite insurgencies and also experienced emergency rule in the mid-1970s. Pakistan underwent political turmoil leading to martial law in 1977 and insurgencies in Balochistan and Sindh.
4. The author's interviews with several Pakistani senior military and civil servants indicate conflicting claims and denials about Pakistan sending their Foreign Minister, Sahibzada Yaqub, to convey a subtle threat, which Yaqub Khan denies having been either tasked or having conveyed. Reports of F-16s being prepared to signal deterrence also remain unverified whether it was a post-event rhetorical claim for domestic political purposes or otherwise.
5. India felt justified in its land grab of Siachin as it was outside the demarcated LOC. The international community saw this crisis as another between India and Pakistan over Kashmir. It was before the start of the South Asian nuclear era, recognized as 1998 and not 1974.

6. The most oft-cited reference is from Bruce Reidal, who was the note taker during the Clinton-Sharif meeting on July 4, 1999. The categorical denial comes from Pervez Musharraf in *In the Line of Fire,* New York: Simon & Schuster, 2006. Also, Lieutenant General Khalid Kidwai, during a briefing tour in the United States in the fall of 2006, repeatedly denied any such preparations. Also see Feroz Hassan Khan's interview with Aziz Haniffa in "Pakistan Did Not Prepare Nuclear Weapons in Kargil Crisis," *India Abroad Weekly Journal*, April 2002.
7. Robert Jervis, The Meaning of Nuclear Revolution: Statecraft and the Prospect of Armageddon, Ithaca, NY: Cornell University Press, 1989.
8. Pakistan took the lead on issues of arms control and disarmament since it had set up a dedicated cell in Army Headquarters in 1994. The author was the first director of this organization, which was later merged with the Strategic Plans Division, Joint Service Headquarters in 1999. See Stephen P. Cohen, *The Pakistan Army*, New York: Oxford University Press, 1998.
9. Statement by Ambassador Munir Akram, Permanent Representative of Pakistan to the United Nations, New York, in the General Debate of the First Committee of the 58th Session of the U.N. General Assembly, October 10, 2003.
10. Strobe Talbott, *Engaging India*, Washington, DC: Brookings Institution Press, 2004.
11. Zafar Iqbal Cheema, "Prospects of Strengthening the CBMs Regime in South Asia," in Pervaiz Iqbal Cheema and Imtiaz Bokhari, eds., *Conflict Resolution and Regional Cooperation in South Asia*, Islamabad, Pakistan: Islamabad Policy Research Institute, 2004, p. 48. "Cheema Cites India Not in Favour of Disarmament," *News India*, November 14, 1987.
12. Pakistan was also confronted on two fronts in the 1980s crises, but its armed forces were not physically involved. It was focused on proxy war against the then Soviet Union.
13. Feroz Hassan Khan, "Nuclear Signaling, Missiles, and Escalation Control in South Asia," in Michael Krepon, Rodney

Jones, and Ziad Haider, eds., *Escalation Control and the Nucleus Option in South Asia*, Washington, DC: Henry L. Stimson Center, 2004, p. 88.

14. Rifaat Hussain, "The India-Pakistan Peace Process," Defence & Security Analysis, Vol. 22, No. 4, 2006, p. 409.
15. Stephen P. Cohen, India: Emerging Power, Washington, DC: Brookings Institution Press, 2001, pp. 204, 209- 211.
16. Scott D. Sagan and Kenneth N. Waltz, *The Spread of Nuclear Weapons: A Debate Renewed*, New York: W. W. Norton & Company, Inc., 2003.
17. Right wing politics in both India and Pakistan generate religious hatred and extremist ideological positions. A ritual cleaning act was performed by Jamait Islami and Shiv Sena respectively after PM Vajpayee's visit to the Pakistan Monument in 1999 and President Musharraf's visit to the Gandhi Memorial in 2001. See Rizwan Zeb and Suba Chandran, "Indo-Pak Conflicts Ripe to Resolve," RCSS Policy Studies, Vol. 34, Colombo, Sri Lanka: Regional Center for Strategic Studies, 2005, p. 23.
18. Walter Ludwig III, "A Cold Start for Hot Wars? Indian Army's New Limited War Doctrine," International Security, Vol. 32, No. 2, Winter 2007/08, pp. 158-190.
19. "Nuclear Safety, Nuclear Stability and Nuclear Strategy in Pakistan," Como, Italy: Landau Network, Centro Volta, January 2002, available from lxmi.mi.infn.it/~landnet/Doc/pakistan.pdf.
20. Barry R. Posen, *Inadvertent Escalation: Conventional War and Nuclear Risks*, Ithaca, NY: Cornell University Press, 1991.
21. Martin Schram, *Avoiding Armageddon: Our Future, Our Choice*, New York: Basic Books, 2003, p. 53.
22. Also author's interview with Martin Schram for PBS Ted Turner Documentaries, PBS series. This was complied in the book, *Avoiding Armageddon*, cited above, which gives identical scenarios extracted from the author's interview, pp. 53-57.
23. India's Prithvi and Pakistan's Hatf series of ballistic missiles, if deployed, may have mixed warheads. Improved surveillance

and intelligence capabilities in both countries will know both deployment sites and launch times; but neither side will ever be certain about the composition of incoming warheads. A launch-to-target time of only a few minutes will reveal the kind of warhead used once the first warhead explodes on target. However, strategic weapons fire exchanges from nuclear-capable delivery systems will inevitably follow, which will leave neither side assured of constant non-nuclear responses through the duration of war. If a conventionally armed warhead launched from a nuclear-capable delivery vehicle targets a nuclear weapon site of the adversary, it is reasonable to believe that a nuclear response would result.

24. The term "never-always" is borrowed from Peter Feaver. See Peter Douglas Feaver, *Guarding the Guardians: Civilian Control of Nuclear Weapons in the United States*, Ithaca, NY: Cornell University Press, 1992.

25. The term "manual override" implies passing the electronic code manually to enable the launching of weapons. In Western jargon, the term "jury-rigged" is often used.

26. Kidwai intervew with Maurizio and Paolo.

27. Manjeet Singh Pardesi, "Deducing India's Grand Strategy of Regional Hegemony from Historical and Conceptual Perspectives," Draft Working Paper No 76, Singapore: Institute of Defence and Strategic Studies, April 2005, available from www.ntu.edu.sg/rsis/publications/workingpapers.asp?selYear=2005.

28. C. Raja Mohan, *Crossing the Rubicon: The Shaping of India's New Foreign Policy*, New Delhi, India: Viking, 2004, p. 156.

29. Peter J Katzenstein, *A World of Regions: Asia and Europe in the American Imperium*, Ithaca, NY: Cornell University Press, 2005, p. 236.

30. The doctrine is named after Prime Minister Indira Gandhi and her son, Rajiv Gandhi, for their security approach in the 1970s and 1980s when India aggressively pursued a policy of assertion with all its neighbours from Sri Lanka to China.

Major military and naval exercises were conducted along the Pakistani and Chinese borders, India flexed its muscles in Sri Lanka with the peace accord of 1987, and it intervened in the Maldives.

31. Andrew Winner and Toshi Yoshihara, "India and Pakistan at the Edge," Survival, Vol. 44, No. 3, Autumn 2002. Also see Rodney Jones, *Conventional Military Imbalance and Strategic Stability in South Asia*, South Asian Strategic Stability Unit, Bradford, UK: University of Bradford, 2005.

32. Zawar Haider Abidi, "Threat Reduction in South Asia," Occasional Paper 49, Washington, DC: Henry L. Stimson Center, 2003.

33. Dalis Dassa Kaye, *Talking to the Enemy: Track Two Diplomacy in the Middle East and South Asia*, Santa Monica, CA: RAND Corporation Study, 2007.

34. The Karachi Agreement of 1949, the Simla Accord of 1972, the Lahore Agreement of 1999, and the Islamabad Accord of 2004 are some of the impressive bilateral accords.

35. An acknowledgement to this effect has been in the Lahore MOU that seeks a mechanism for the implementation of existing CBMs.

36. Karan R. Sawny, "The Prospects for Building a Peace Process Between India and Pakistan," in Cheema and Bokhari, eds., *Conflict Resolution and Regional Cooperation in South Asia*, p. 32-40.

37. See, for example, the statement by Ambassador Munir Akram in the general debate of the first committee of the 58th session of the U.N. General Assembly, New York, October 10, 2003.

38. At the time of this writing in September 2008, there is an unprecedented tension between United States and Pakistan. Pakistan has strongly protested U.S. Special Forces' cross-border incursions and open statements by U.S. policymakers to expand the war into Pakistani territory.

39. India and Pakistan should engage in the three areas bilaterally. The initial U.S. role should be to act as a catalyst and honest broker between allies.

40. David Albright, "Securing Pakistan's Nuclear Weapons Complex: Thought-Piece for the South Asia Working Group," paper presented at the Stanley Foundation conference on "US Strategies for Regional Security," Airlie Conference Center, Warrenton, VA, October 25-27, 2001.
41. Dingli Shen, "Upsetting a Delicate Balance," Bulletin of the Atomic Scientists, Vol. 63, No. 4, July 1, 2007, p. 37.
42. Mohan, pp. 155-156.

Chapter - 3

Is Nuclear Power Pakistan's Best Energy Investment?

Assessing Pakistan's Electricity Situation

—John Stephenson and Peter Tynan

Introduction

The drive for civil nuclear power has resurged around the globe, often under the banner of finding a clean energy alternative to meet growth objectives. Countries like India, Saudi Arabia, United Arab Emirates (UAE), Turkey, Egypt, and Jordan, among others, have all proclaimed a desire for nuclear power generation. Proponents argue that nuclear energy promotes economic development and reduces reliance on foreign sources of energy in a manner that is climate-change friendly due to the lack of carbon emissions.

Similarly, Pakistan has pushed for nuclear power generation using many of the same arguments. Advocates for this initiative have underscored the recent congressional approval of the U.S.-India Civil Nuclear Cooperation agreement, which provides India with access to nuclear equipment and components from Western suppliers. As Pakistan's Prime Minister Yousaf Raza Gilani stated: "Now Pakistan also has the right to demand a civilian nuclear agreement with America. We want there to be no discrimination. Pakistan will also strive for a nuclear deal, and we think they will have to accommodate us."[1] A critical question, however, is whether nuclear power is necessary and vital to economic development in a climate-change friendly manner.

This analysis looks at the economic and resource arguments for nuclear power through 2030 to evaluate whether nuclear power is necessary to meet Pakistan's energy expectations. First, the analysis evaluates the assertion that nuclear energy is vital to meet economic development goals. Second, this chapter analyzes the claim that global carbon emissions will be reduced by such an amount as to make salient the argument for increased Pakistani nuclear power generation capability. Finally, it evaluates whether development of nuclear energy would significantly reduce Pakistan's reliance on foreign energy sources. The framework used to evaluate resource options for electricity development (see Figure 1) includes looking at the total potential capacity, the likely pace of development of different technologies, the relative costs of those options, and the environmental issues and trade-offs inherent with each option.

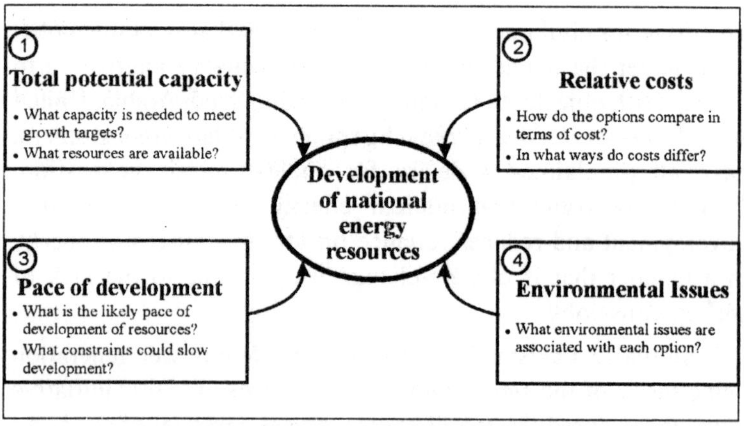

Figure 1. *Analytical Framework.*

This analysis concludes that nuclear power does not meet the expectations laid out by advocates for its development in Pakistan through 2030. Even under Pakistan's most ambitious growth plans, nuclear energy will continue to contribute a marginal amount of electricity to meet the country's economic goals. Furthermore, with Pakistan's considerable potential of untapped renewable resources, the country has numerous options other than nuclear to meet its development needs. In terms of reductions of carbon emissions, it should be noted that Pakistan currently represents only about 0.4 percent of global emissions. Certainly, while all emissions reductions are necessary, such reductions need to be pursued within the context of other risks, whether from deferred economic development or proliferation of sensitive technologies. Finally, given the sources of energy supplying Pakistan's electricity generation, a significant proportion of which is based on natural gas, Pakistan could reduce its reliance on foreign sources of energy by developing nuclear. However, nuclear in the best case scenarios will provide a limited amount of electricity, and the predominant foreign sources of energy still emit carbon. As such, the route to developing Pakistan's considerable renewable resources can achieve the dual goals on carbon reduction and enhanced self-reliance.

Background: Current and Future Needs

The primary sources of Pakistan's electricity are natural gas, hydro, and oil/diesel generation (see Figure 2). The total generation capacity of Pakistan in 2005 was 19.5 gigawatts (GW) and consisted of approximately 50 percent from natural gas, 30 percent from hydro power, and 16 percent from oil/diesel. Nuclear power's current contribution of electricity generation is 3 percent, while the contribution from coal is only 0.2 percent. Notably, renewable energy resources did not contribute to Pakistan's generation capacity in any meaningful way in 2005.

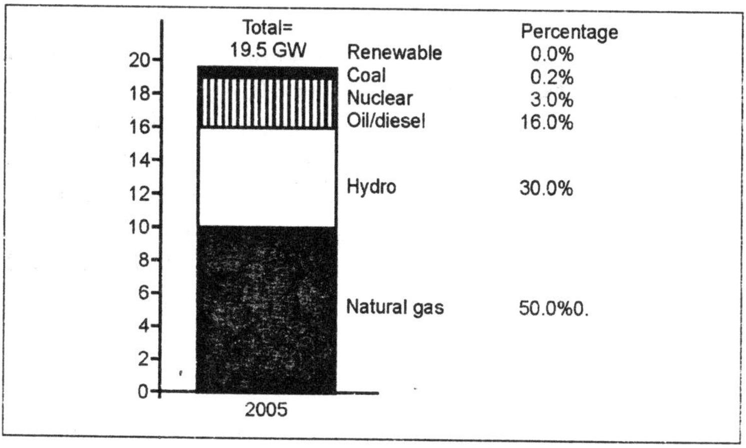

Figure 2. *Pakistan's Current Electricity Generation*

Capacity, 2005 (GW).[2] Pakistan's current electricity generation capacity also does not meet the current demand, creating significant shortfalls. The country is presently experiencing supply deficits during peak demand periods and the variability of water supply contributes to deficits given the large reliance on hydropower.[3] Nearly half of the population is also estimated to lack connection to the electricity grid, and load shedding has also become necessary in some areas.[4] Some estimates suggest that the grid system requires approximately two additional GW to cover peak demand with an adequate degree of reliability.

Compounding the challenges for meeting current demand, Pakistan's generation capacity requirements are expected to increase significantly through 2030 (see Figure 3). Forecasts for this growth rate vary and are generally tied to gross domestic product (GDP) expansion, which represents the energy intensity of economic growth. The Government of Pakistan estimates are based on an 8 percent GDP growth rate and a corresponding 9 percent generation capacity growth rate, thereby requiring 163 GW of generation capacity by 2030. However, the historical generation capacity growth rate from 1980-2005 was roughly 7.1 percent, and, if this trend continues, the capacity by 2030 would likely be 108 GSs. The actual generation capacity developed by 2030 will likely be somewhere in between these two ranges. However, even assuming a stronger GDP growth rate of 8.5 percent, thereby exceeding the Government of Pakistan projections, the need would be roughly 193 GSs. While the energy intensity varies and tends to decrease as an economy develops, the estimates of generation capacity present a conservative range against which to test the need for specific supply options. Considering the recent global financial and economic downturn, Pakistan's GDP growth rate could be significantly constrained, which could also create a concurrent reduction in the need for generation capacity.

Total Potential Capacity

Despite the considerable power generation requirement needed by 2030, Pakistan has a wide breadth of potential sources to meet this future demand. In comparing the potential supply of resources with the generation capacity needed by 2030, this analysis uses both low- and high-end ranges based on various projections of GDP growth. As discussed above, the estimated generation capacity required by 2030 will be between 108 GWs and 193 GWs (shown in Figure 3). For each potential supply, the analysis also uses low and high estimates for the development through 2030.

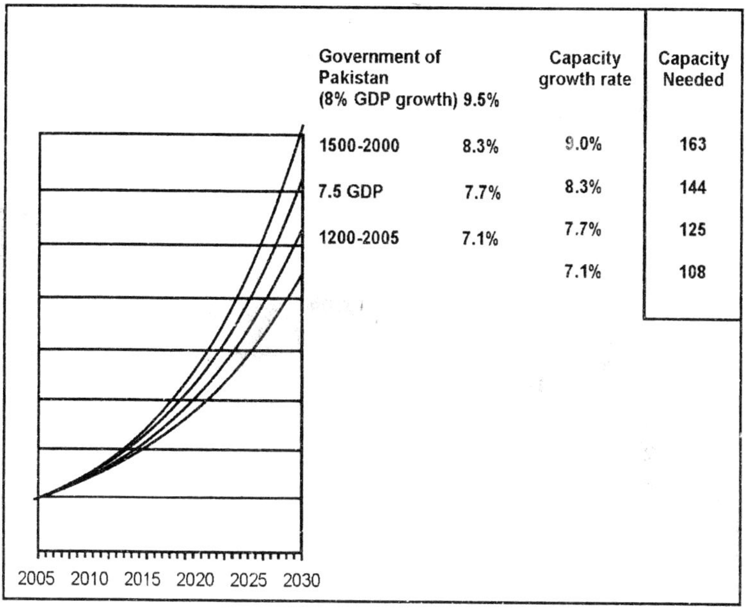

Figure 3. *Projections for Pakistan's Generation Capacity Requirements, 2006-2030.*[5]

The key finding is that the potential supply of resources should be capable of meeting both low and high estimates for generation capacity needs, although requiring a portfolio approach. The available and likely resources consist of a broad range of supply options involving considerable development of the traditional supply sources of natural gas, hydro, and coal. While indigenous natural gas supplies are expected to dwindle, the Government of Pakistan has committed itself to investing in accessing external sources through pipelines.[6] In terms of coal, Pakistan has approximately 185 billion tons of reserves, even with the anticipated increase of approximately 2.2 GW of coal-generated electricity by April 2009. Renewable energy resources offer significant potential even in the low and medium scenarios, which do not maximize the utilization of these resources, thereby leaving additional potential for well beyond 2030. Energy efficiency options are also likely to be a meaningful contributor to the variety of resources by 2030, offering more potential than that of nuclear power.

Notably, even if the development of nuclear power meets high estimates, it is unlikely to constitute a significant contribution to the overall supply. Currently, Pakistan has two nuclear power plants (Chashma-1 and Kanupp) which generate 300 megawatts (MW) and 125 MW, respectively. Pakistan's third nuclear power plant, Chashma-2, was expected to be completed by 2009 and will be capable of generating 325 MW. The Government of Pakistan estimates suggest a 13 percent growth rate (see Figure 4) which would yield approximately 6-8 GW of nuclear power generation.[7] This represents only about 3-6 percent of the electricity generated in 2030. If those high estimates are not met but instead nuclear power generation grows at a fast yet more reasonable pace of 8 percent, the total nuclear power generation would be roughly 2.8 GW. This would constitute only 1-3 percent of the total generation capacity by 2030.

Pakistan has considerable solar potential that rivals many other regions of the world. The solarization of the country averages 5.2 kwh/m^2 and nearly half of the country shows economic viability for solar power generation. Few regions, aside from the Sahara, offer better solar potential in the world. Both solar photovoltaic and concentrated solar thermal technologies are becoming increasingly cost effective and commercialized, offering a considerable opportunity for this untapped resource in Pakistan. Although estimates for the total potential generation capacity from solar vary, a reasonable estimate is 70 GW.[8]

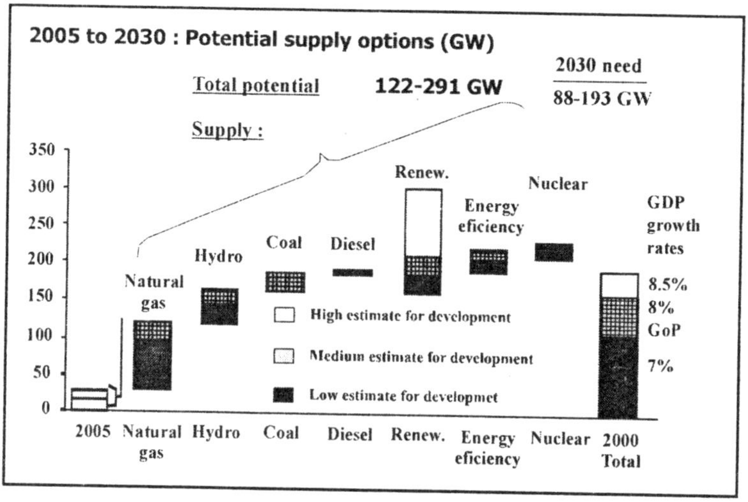

Figure 4. *Pakistan's Potential Supply Options for Electricity Generation, 2005-2030.*[9]

The opportunity for wind power generation is also quite significant in Pakistan, at approximately 50 GW of potential generation capacity and a target of 9.7 GW by 2030.

The AEDB is facilitating favourable rental terms for developers, and numerous letters of intent have been signed, with the target of generating 9.7 GW by 2030.[10] The National Transmission and Despatch Company (NTDC) is constructing new transmission lines to bring the power to markets, and at least two urban hubs, Karachi and Hyderabad, are nearby.[11] The potential of an estimated 50 GW of generation capacity suggests that ample wind capacity will still be available long after 2030.

In addition to solar and wind, other promising renewable energy sources exist for Pakistan to develop more fully. For instance, sugar mills in the country use bagasse for cogeneration purposes, and the Government of Pakistan has recently enabled them to sell surplus electricity back to the grid. Other such biomass, biogas, waste-to-electricity, and biofuels could also meaningfully contribute to the energy and electricity supply in Pakistan. The estimate for waste-to-electricity alone is approximately 500 MW per major city.[12]

Given the split between rural and urban populations, decentralized generation sources could also make considerable sense for development in Pakistan. By some estimates, roughly 70 percent of the population lives in rural villages,[13] with nearly half the population lacking a grid connection. With the costs of transmission and distribution, it is often uneconomical to connect these populations to the grid. As such, a centralized power generation source, like nuclear, may not serve to increase electrification rates across the country. Instead, decentralized wind and solar generation can often serve these populations better, and many small scale projects have already been developed throughout the country. The other concern is to have sufficient baseload generation, for which nuclear is normally used. While some renewable technologies raise concerns of intermittency, new technologies are being commercially developed to provide storage and enable use for baseload generation, especially as seen with concentrated solar thermal. And given the small share of nuclear power in the overall generation capacity mix by 2030, other options like hydro will provide significant baseload generation.

Another critical opportunity for meeting Pakistan's electricity needs will be in energy efficiency, or negawatts, which even with conservative estimates will amount to more than nuclear power generation. Energy efficiency efforts can tackle a number of key areas in electricity production and consumption. They can include improving demand or efficiencies, such as switching to improved lights and energy efficiency appliances. Industrial production of goods can similarly be improved to generate considerable negawatts. Electricity generation itself can also be made more efficient, particularly with thermal generation, through equipment upgrades. Finally, transmission and distribution losses, traditionally quite high in developing countries due to technical losses and theft, can be improved for significant savings. Currently, Pakistan's transmission and distribution losses are estimated at approximately 26.5 to 30 percent.[14] The Government of Pakistan set the goal of reducing these by 5 percent by 2010, which could create approximately 8 GW of negawatts cumulatively by 2030.[15] Committing to another 5 percent reduction in transmission losses would double this to roughly 16 GW. In terms of estimating negawatts, this analysis remains quite conservative, having only reflected the potential savings from improving transmission and distribution losses. If demand efficiencies had been incorporated, these estimates could be considerably higher. Regardless, the potential improvements in transmission and distribution losses alone would outpace nuclear power generation by 1.5-3 times.

Pace of Development

The likely pace of development of various supply options will be especially important for Pakistan and current projections significantly outpace historical development. Government projections often suggest that generation options will develop much more rapidly than historical progress suggests, and projections of nuclear development are no exception. In fact, the projections of nuclear development in Pakistan are predicated on attaining a development trajectory that very few countries in the world have been able to attain.

When projecting likely development of electricity generation sources, it is important to look first at the historical development of various options (see Figure 5). In the case of Pakistan, the development of thermal and nonconventional energy sources (NCES) has risen the fastest over a 25-year period, at approximately an 8 percent growth rate. However, recently (from 2000-05), installed plant capacity from these sources has stagnated at 0 percent growth, while hydro power, at 6 percent, has maintained a consistent growth rate over the entire 25-year time period. Nuclear grew the slowest over this period, at 5 percent from 1980-2005.

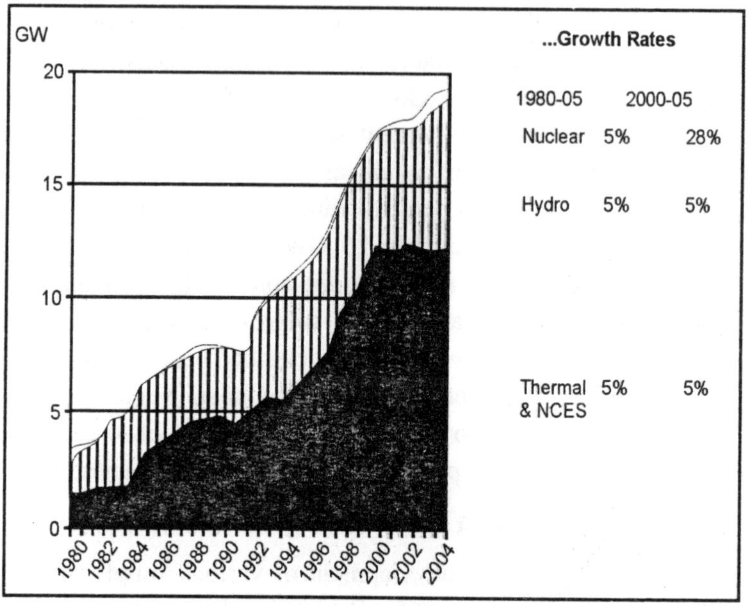

Figure 5. *Historical Development of Electricity*

Recent high growth rates of 28 percent from 2000-05 reflect the small number of nuclear power plantsoverall. With two plants online and a third scheduled to go live in 2009, each additional plant represents a significant percentage of the total. Overall, generation capacity grew by 7 percent from 1980-2005 and by only 2 percent from 2000-05.[16]

The projections for the various supply options are almost uniformly ambitious, but especially so for nuclear. From 2005 to 2030, it is expected that nuclear generation will increase at a growth rate of 13 percent.

Other supply options have similarly high estimated growth rates, such as natural gas at 9 percent, hydro at 10 percent, coal at 13 percent, and renewable energy at 14 percent.

Nearly all of these supply options will undoubtedly face challenges in attaining such growth targets, but fewer challenges are likely be met by those options that face lower barriers in the form of capital intensity, political will, and ready availability of supplies and technology.Generation, 1980-2005 (GW).[17] Because nuclear faces immense challenges in terms of capital intensity and accessibility of supplies and technology, the growth rates implied for nuclear development suggest the attainment of targets that very few countries in the world have been able to achieve. However, as a nonsignatory to the Non-Proliferation Treaty, there are international embargoes on the transfer of such technology to Pakistan. China is currently the only supplier of nuclear power plants and components to Pakistan, but, to meet the projections, Pakistan would require access to advanced nuclear supplies and technologies from Western countries.[18] Such constraints raise particular questions around nuclear development, where governments are especially prone to overestimate their ability to develop such resources and install generating capacity. (See Figure 6.)

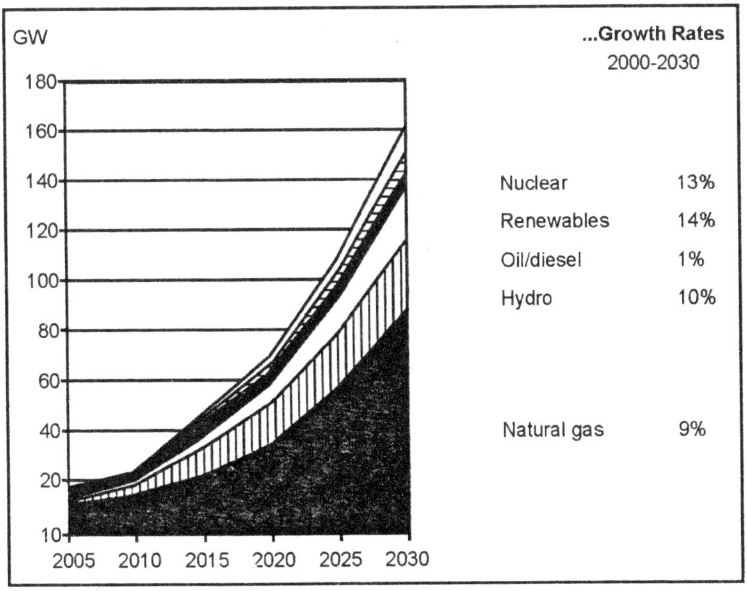

Figure 6. *Projected Development of Installed Plant Capacity, 2005-2030 (GW).*[19]

Globally, the historical data of nuclear power development suggests that few countries have been able to achieve and maintain a consistently high growth rate for nuclear development as per Pakistan's estimates. South Korea comes the closest to reaching the trajectory and sustainability of nuclear power generation with a 14.3 percent compound annual growth rate (CAGR) over the 15 years from 1980 to 2005. The United States and France both had much faster growth from 1980 until approximately the early 1990s (at 7 percent and 14 percent respectively), but their nuclear development programmes have since leveled off.[20] By contrast, India has only attained a 4.9 percent growth rate for its nuclear development.[21] For Pakistan to meet its own nuclear power development estimates over the next 30 years, it would have to emulate or surpass the efforts of countries like South Korea, France, or the United States. (See Figure 7.)

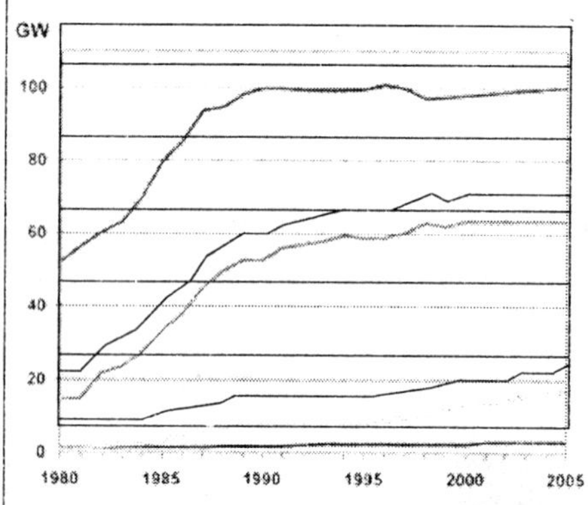

	Installed GW		1980-2005 CAGR
US	52	100	2.7%
France	14	63	6.1%
South Korea	0.59	17	14.3%
India	0.9	2.8	4.9%

Figure 7. *Historical Development of Nuclear Power in the U.S., France, South Korea, and India, 1980-2005 (GW).*[22]

Nuclear development also requires considerable coordination between the private and public sectors, requiring rather strong government effectiveness, regulatory quality, and control of corruption since nuclear projects require large capital expenditures. Relative to countries such as the United States, France, and

South Korea that have successfully developed nuclear power generation at impressive growth rates, Pakistan's measure on these governance indicators is significantly lower. (See Figure 8.) Although these metrics are general governance indicators, the successful implementation of a nuclear power development policy would presumably require even greater government effectiveness, regulatory skills, and control of corruption than ordinary largescale infrastructure projects. While a lower rating in government effectiveness may suggest a country is less able to orchestrate the necessary level of coordination to get a project initiated and complete, a lower rating in regulatory quality would suggest potential lapses in security and safety, and less control of corruption would suggest that sensitive materials may be more prone to illicit sale and trade. Corruption also matters considerably in terms of the financing of largescale infrastructure projects. The "corruption tax" on a large project can significantly balloon costs and delay completion. These discrepancies in governance indicators would suggest that the nuclear generation growth rates targeted by the Government of Pakistan may not be achievable.

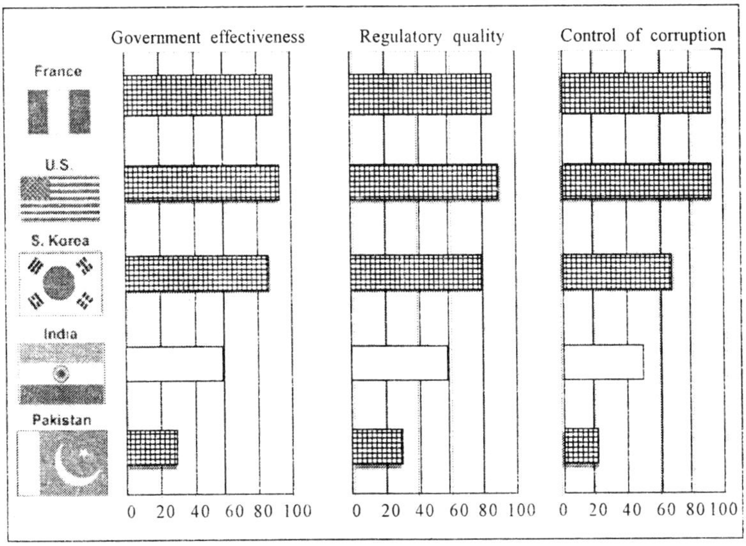

Figure 8. *Governance Indicator Comparison, 2007.*[23]

At the same time, the regulatory and policy environment for renewable energy development, including wind and solar power, is being increasingly strengthened and geared towards enhancing and accelerating development. Legislation that has been passed includes sales tax, income tax, and customs duty exemptions for imported plants, machinery, and equipment for renewable energy power generation.[24] Further incentives for private sector development of wind power even includes "Wind Risk Coverage," which covers the risk of wind speed variability, making the power purchaser (the Government of Pakistan) absorb the risk of such variability.[25] The AEDB continues to lobby aggressively for investments and, in the case of wind, roughly 93 letters of intent have already been signed for development.[26] This push has benefited from foreign assistance, such as support from the U.S. National Renewable Energy Laboratories under a 2007 U.S. Agency for International Development (USAID) assistance programme.[27]

Relative Costs

The likely development of various supply options is influenced by a number of factors, including the relative costs of those options (See Figure 9). Estimates of the relative costs of different supply options vary widely. By far, the lowest cost options are coal and hydro, while some of the most expensive options are solar photovoltaic and solar thermal. Local costs of supply options can vary considerably, and Pakistan specific estimates suggest nuclear energy could be on the high-end of the range, at roughly $0.057 cents per kilowatt-hour (kWh).

It is also important to note the trend for the cost curves of renewable energy technologies. Wind has led the way in becoming economically viable, and solar is expected to follow suit in the medium term. The price of concentrated solar power has dropped at a faster rate than solar photovoltaic (PV), but recent advances in solar PV technology also suggest increased commercial viability.

Furthermore, for many nongrid connected Pakistanis, the trade-off is not necessarily between cheap sources of electricity

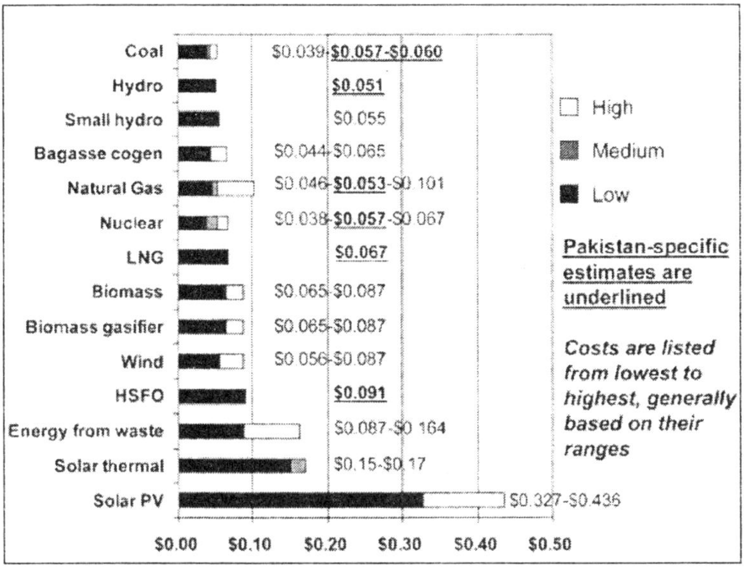

Figure 9. *Relative Costs of Various Resource Options.*[28]

or renewable options. Rather, it lies in which resources can, or will, be developed in the near-, medium-, and long-term. A more expensive option per kilowatt-hour, like solar or wind, may have lower up-front costs and not rely on the central government to invest in infrastructure requirements for transmission and distribution. One significant benefit of renewable energy technologies like wind and solar, however, is that they both have predictable (i.e., zero) fuel costs and can also be expanded incrementally. Wind and solar farms can largely be built in stages, with the first phases of installation becoming immediately productive, while a fractional buildout of a nuclear facility cannot produce electricity.

It is also important to note the potential ramifications of the current global financial crisis. As access to capital becomes constrained, it will likely become more difficult to finance large scale investments like nuclear, especially where the production of electricity and generation of cash flows comes much later. Less capital-intensive projects that can be built-out incrementally are more likely to be favoured and will be used to meet electricity demand that itself is likely to be reduced due to economic growth constraints.

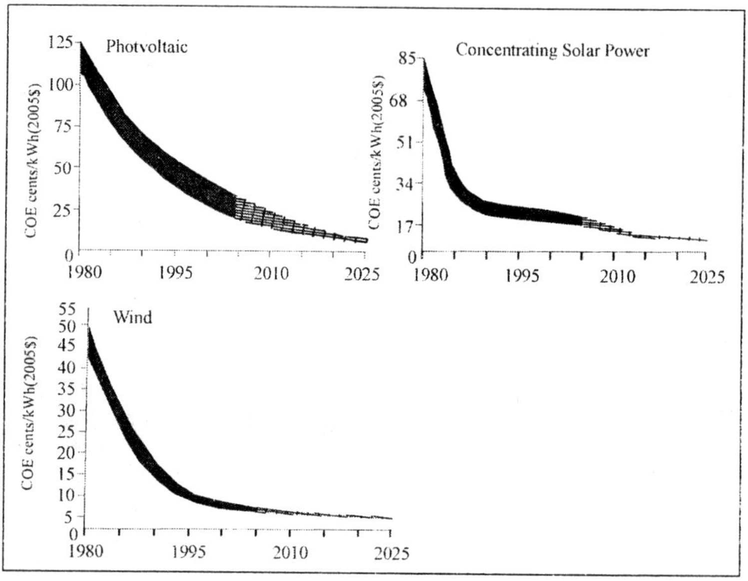

Figure 10. *Cost Curve for Solar Photovoltaic, Concentrating Solar Power, and Wind, 1980-2025.*[29]

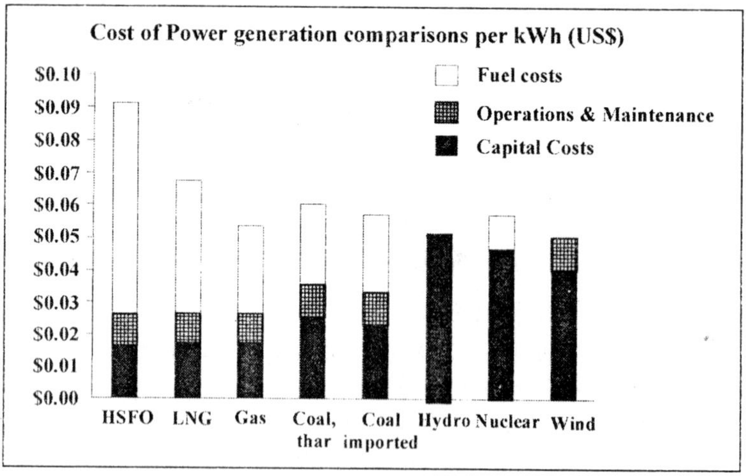

Figure 11. *Cost Structure Comparison of Various Supply Options (per kWh, US$).*[30]

In the end, it is important to compare the relative costs of different supply options, but meeting Pakistan's electricity needs will require a portfolio strategy. No single option, no matter how attractive from a cost perspective, can meet the full need by 2030. Numerous options need to be pursued, leveraging the strengths and mitigating the risks associated with each.

ENVIRONMENTAL ISSUES

Given the very real risks of climate change, it is vital to consider environmental issues when evaluating electricity supply options in any region of the world. Nuclear is often judged against a "clean" generation technology due to the lack of carbon emitted during electricity generation. While this is true, renewable energy technologies are equally climate-change friendly and are not accompanied by the problems associated with long-lasting radioactive spent fuel and its transportation, storage, and disposal.

It is also important to look at the sources of carbon emissions by country to determine the appropriate intervention to reduce those emissions. In Pakistan, a significant amount of carbon emissions comes from petroleum which serves transportation needs

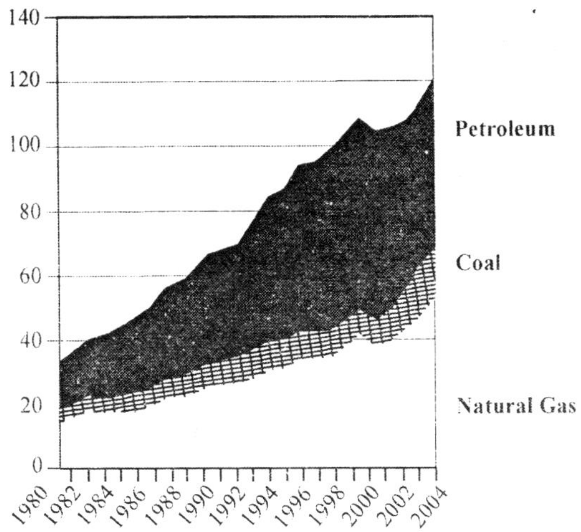

Figure 12. *Pakistan's CO_2 Emission by Source, 1980-2005 (million metric tons).*[31]

and would not be offset by switching to electricity generation resources, at least until electric cars are widespread in Pakistan. A promising trend in Pakistan's transportation sector, however, is the increased use of compressed natural gas for transportation.[34] Also, while a significant amount of Pakistan's emissions come from natural gas (including for electricity generation), natural gas produces just about half the emissions of coal.

Finally, while all emissions reductions contribute to addressing the issue of climate change, Pakistan's emissions should be considered in context when weighing the attractiveness of other options involving different types of risks. In 2005, Pakistan produced just 0.4 percent of total global carbon emissions. By comparison, Pakistan produces only 0.77 metric tons per capita versus 20.14 metric tons per capita in the United States. As such, the degree of the carbon emissions problem in Pakistan may not outweigh other the risks associated with nuclear power generation. This is especially true when considering the ample renewable energy potential in Pakistan, the benefits of decentralized power generation in the country, the decreasing costs of renewable energy sources, and the lack of fuel risks attaching to renewable energy sources (both in terms of price volatility and spent-fuel risks).

CONCLUSION

Numerous countries, including Pakistan, are pushing to develop nuclear power generation capacity. These countries often highlight the requirements of economic development to increase their electricity generation. In a carbon constrained world with increasing global awareness of the risks of climate change, nuclear power is judged as a clean and efficient way to meet economic development objectives while limiting carbon emissions. Furthermore, nuclear power is often seen as a means of ensuring greater selfreliance and independence from petroleum imports from unstable neighbors or regions. With the recent approval of the U.S.-India Civil Nuclear Cooperation agreement, Pakistan is also calling for access to nuclear equipment and supplies from

Western sources as a measure of fairness and support for its economic development.

However, in the case of Pakistan, the promises of nuclear power generation are largely exaggerated through 2030. While it remains true that Pakistan currently has an electricity generation capacity shortage and will need considerably more capacity by 2030, there is ample potential supply from numerous other sources. Traditional sources such as natural gas and hydro will continue to be important for Pakistan, but increasingly, the potential of renewable energy will be harnessed. Pakistan is extremely well-endowed not only with large-scale hydro, but also worldleading solar and wind resources. The government has recognized this by establishing the AEDB, and has increased the amount of investments in this sector.

With a portfolio approach encompassing traditional and renewable energy sources along with energy efficiency measures, Pakistan can meet its electricity needs through 2030 if it chooses to forego nuclear power development. The role of nuclear in the mix of electricity generation sources by 2030 is not vital. First, the estimates for nuclear development are quite ambitious and rest on the assumption that Pakistan could replicate the development trajectory of the United States, France, and South Korea. Second, nuclear development requires significant private and public sector coordination resting on a solid foundation of government effectiveness, regulatory quality, and control over corruption. Compared with those countries that have successfully developed nuclear power in the past, Pakistan falls short of these metrics. Finally, even if the high estimates are achieved by Pakistan, the resulting contribution would represent only 3-6 percent of total electricity generation capacity. Furthermore, Pakistan's overall contribution to global carbon emissions remains miniscule at 0.4 percent, so substitution through an aggressive nuclear energy programme does not suggest meaningful progress on the climate change agenda.

Endnotes

1. Damien McElroy and Rahul Bedi, "Pakistan Demands Rights to Nuclear Power after India Deal Is Sealed," *Telegraph*, October 2, 2008, available from *www.telegraph.co.uk/news/3122690/*Pakistan-demands-rights-to-nuclear-power-after-India-deal-is-sealed. *html.*
2. *Pakistan in the 21st Century: Vision 2030*. Islamabad, Pakistan: Government of Pakistan, 2007; "Pakistan Approves 25-year Energy Security Plan, 2005," *Pakistan Times*, 2005.
3. Vladislav Vucetic and Venkataraman Krishnaswamy, "Development of Electricity Trade in Central Asia-South Asia Region," available from *mea.gov.in/srec/internalpages/drafteti.pdf.*
4. Energy Information Administration, available from *www.eia.doe.gov/cabs/Pakistan/Electricity.html.*
5. *Medium Term Development Framework and Vision 2030*; *Pakistan Statistical Yearbook 2007*, Islamabad: Government of Pakistan, Federal Bureau of Statistics, 2007; *Pakistan 25-year Energy Security Plan 2005*, Dalberg analysis, *Pakistan Times*, 2005.
6. *Energy Information Administration Country Analysis Briefs-Pakistan* (EIA Pakistan Country Analysis Brief), available from *www.eia.doe.gov/cabs/Pakistan/Profile.html.*
7. *Pakistan In the 21st Century; Pakistan Times*, 2005; *Pakistan 25-Year Energy Security Plan, 2005*; Angelica Wasielke, ed., *Energy-Policy Framework Conditions for Electricity Markets and Renewable Energies*, Eschborn, Germany: ETZ, 2004; *23 Country Analysis*, 2007, Pakistan Chapter, pp. 339-359. Notes: Nuclear represents the projected 13 percent growth rate for the high estimate and an 8 percent growth rate which is still very high compared with other countries histories of nuclear development.
8. Wasielke, ed., p. 11.
9. *Pakistan 25-year Energy Security Plan, 2005*, *Pakistan Times*, 2005.
10. "Resource Potential of Wind Project," Islamabad, Pakistan: AEDB, available from *www.aedb.org/res_potential.php.*

11. Wasielke, ed., p. 9.
12. Wasielke, ed., p. 10.
13. "Resource Potential of Solar Protovoltaic Project, Islamabad, Pakistan: AEDB, available from *www.aedb.org/respotential_pv.php*.
14. Estimates of 30 percent by the Energy Information Administration, *www.eia.doe.gov/cabs/Pakistan/Electricity.html*.
15. *Mid Term Review of Medium Term Development Framework Framework, 2005-2010*, Islamabad: Government of Pakistan, Planning Commission, 2008, p. 84.
16. Energy Information Administration.
17. *Ibid.*
18. "Nuclear Power Programme of Pakistan," paper by Zia H. Siddiqui, Tariq Mahmud, and Ghulam R. Athar, presented to World Nuclear Association Symposium, 2006.
19. "Pakistan 25-year Energy Security Plan," *Pakistan Times*, 2005.
20. DoE Energy Information Administration report on Pakistan.
21. *Ibid.*
22. *Ibid.* Note: The fastest growing installed nuclear capacities are in South Korea (14.3 percent), Spain (8.1 percent), and France (6.1 percent).
23. World Bank Governance Indicators, available from *info.worldbank.org/governance/wgi/index.asp*.
24. AEDB, available from *www.aedb.org*.
25. *Ibid.*
26. "Resource Potential of Wind Project," AEDB, available from *www.aedb.org/res_potential.php*.
27. *Ibid.*
28. Sources: Pakistan-specific estimates (others are benchmarks, with some estimates from elsewhere in South Asia) from Mukhtar Ahmed. "Meeting Pakistan's Energy Needs," Woodrow Wilson Report, *Fueling the Future: Meeting Pakistan's Energy Needs in the 21st Century*; Energy Information Administration. *Annual Energy Outlook*, DOE/

EIA-0383(2006), Washington, DC, February 2006; Government of India, Planning Commission, *Integrated Energy Policy: Report of the Expert Committee*, New Delhi, India: Government of India, August 2006; David G. Victor, "The India Nuclear Deal: Implications for Global Climate Change," Testimony before the U.S. Senate Committee on Energy and Natural Resources, July 18, 2006, available from *www.cfr.org/publication/11123/india_nuclear_ deal.html*. Notes: For the levelized cost comparison of coal, natural gas (advanced combined cycle), the low estimate for wind, and nuclear, the cost comparison is for U.S. plants that would come online in 2015. For the nuclear generation estimates by David Victor, for light water reactors: the lowest at 3.8 U.S. cents comes from Bharadwaj, Anshu; Rahul Tongia, and V. S. Arunachalam, "Whither Nuclear Power?" *Economic and Political Weekly*, Vol. 41, No. 12, 2006, pp. 1203-1212. The medium cost of 4.2 cents per kWh and 6.7 cents per kWh come from *The Future of Nuclear Power: an Interdisciplinary Study*, Cambridge, MA: MIT Press, 2003. Using the US DOE's levelized costs and incorporating the fact that Indian fuel is 2-3 times costlier, a cost of 6.6 cents per kWh is estimated. "HSFO" is heavy fuel oil.

29. U.S. National Renewable Energy Laboratory.

30. Mukhtar Ahmed, "Meeting Pakistan's Energy Needs." U.S. DOE levelized cost calculations.

31. Energy Information Administration.

Chapter - 4

Pakistan's Economy:
Its Performance, Present Situation and Prospects

—Shahid Javed Burki

Introduction

Pakistan currently faces a grim economic situation. There is likely to be a sharp reduction in the rate of economic growth, an unprecedented increase in the rate of inflation, a significant increase in the incidence of poverty, and a widening in the already large regional income gap while the fiscal and balance of payments gaps increase to unsustainable levels. The country has been through many crises before, but the one that it is currently experiencing is uniquely severe. Should the economic situation continue to deteriorate, the country could be plunged into social and economic chaos that would affect the rest of the world. Pakistan is already considered to be the center of Islamic extremism, so how should it tackle this situation?

In an article published by *Dawn* on July 22, 2008,[1] I suggested that Pakistan should not approach the donor community with a begging bowl in hand and ask for help to resolve the current economic crisis. I did not advocate going to the International Monetary Fund (IMF) for support since that would compromise the effort to keep the economy growing. This is what the country did in 1999 and gave up growth in favour of stabilization. In an effort to increase growth, the Musharraf administration loosened fiscal and monetary controls over the economy and laid the foundation of the current crisis. It is not good for the economy to go through such deceleration and acceleration in growth; repeated shifts are destabilizing, and it would not be prudent to send the economy through such a cycle again.

Instead, I suggested that the country should seek help on the basis of a well-thought-out programme of economic reform and focus on bringing about structural changes that have long been postponed. An important structural change would be to make the economy less dependent on external help for sustaining growth. This will take time, but the process must begin.

By initiating a programme of structural reform, the country may be able to secure long-term finance, perhaps as much as $40 to $50 billion for a 5-year period. Financing should be equally shared between the donor community and Pakistan, with the donors requested to front-load the effort with $20 to $25 billion provided in the first 2 to 3 years, and the Pakistani government providing a matching amount at the end of the programme period. However, the Pakistani authorities should clearly and persuasively describe how it would raise this amount of money.

I cannot tell whether my thinking influenced the policymakers in Islamabad, but I am struck by two developments. First, Pakistani Finance Minister Naveed Qamar made a statement on September 19, 2008, that his government had no intention of going to the IMF for support and that instead it would develop its own package of reform. To reinforce the point, he announced the withdrawal of a number of consumer subsidies that weighed heavily on the federal budget. Secondly, President Asif Ali Zardari, while on a visit to New York to attend the opening session of the United Nations (UN) General Assembly a week later, met with a group of donors he called the "Friends of Pakistan." The group promised support but did not elaborate a plan as to how that would be delivered. This is the situation today as the country continues to diminish the respectable level of foreign currency reserves it had built up over the last 8 years. Within a few months, it will run out of reserves and may have to default on its foreign obligations.

Soliciting donations is only half of the solution to the mounting crisis. The second half of the effort would be to develop a strategy to reassure the community of donors that the new leaders are up to the task of bringing the country out of the stiffest challenge it has faced in its history. Such an effort will need a great deal of thought, the full commitment on the part of the leadership, and public support. It will also need the creation and development of the institutional infrastructure that is needed to support a far-reaching programme of economic and social restructuring.

Time is running out for Pakistan. The approach to the donor community for help should include the presentation of a well-developed, carefully budgeted, and implementable programme of economic change and reform. We need to dispense with the begging bowl approach and adopt one that makes a selected number of countries Pakistan's economic partners rather than providers of charity. At this point, it would be useful to provide a brief historical overview of Pakistan's economic history before examining the current problems the country faces and the policies it could adopt to resolve them.

An Overview of Pakistan's Economic History[2]

Pakistan's performance has been fairly impressive in terms of economic growth and development over the last 60 years. If we construct three indices: growth in population, increase in gross domestic product (GDP), and increase in per capita income for the past 60 years (See Table 1), we notice reasonable progress. While the population increased more than five times, from 30 million in 1947 to over 162 million now, both GDP (which increased 18 times) and per capita income (which increased more than 4 times) also grew appreciably.

However, progress was neither gradual nor even. There were three periods of high growth (1958-69, 1977-88, and 2002-07)—27 years out of 61 years—during which GDP increased by an average of 6.2 percent a year. (See Table 2.) This means that

Table 1. Indexes of Growth in GDP, Population, and Income Per Capita.

Year	GDP	Population	GDP/N
1947	100	100	100
1958	134	122	110
1969	257	156	166
1971	288	165	176
1977	362	198	185
1988	725	277	266
1999	1,201	371	330
2007	1,816	445	418

Source: *Calculated from Government of Pakistan, Pakistan Economic Survey, various years, Islamabad, Pakistan.*

Table 2. Economic Performance during Various Political Periods (Percent).

Year	GDP	Population Growth Rate	GDP Growth Rate
			Capita Increase
1947-58	2.7	1.8	0.9
1958-69	6.1	2.3	3.8
1969-71	5.8	2.8	3.0
1971-77	3.9	3.1	0.8
1977-88	6.5	3.1	3.4
1988-99	4.7	2.7	2.0
1999-2002	5.3	2.3	3.0
2002-07	7.0	1.8	5.2

Source: *Government of Pakistan, Pakistan Economic Survey, various years, Islamabad, Pakistan.*

one-half of the GDP expansion came in those 27 years. Before identifying the reasons for the booms and busts of the Pakistani economy, it would be instructive to compare the country's performance with that of its neighbour, India.

Comparing the performance of the Indian and Pakistani economies in terms of the growth in GDP highlights one important conclusion. The acceleration in the rate of growth of India since the mid-1980s represents a paradigm shift. Between 1947-87, the Indian economy registered what Raj Krishna, an Indian economist, famously called the "Hindu rate of growth." This was about 3.5 percent a year and represents a relatively low level of increase in per capita income. Since the mid-1980s, the Indian economy has been growing annually at rates between 6 and 9 percent. It is fair to conclude that the Indian policymakers were able to put the economy through a deep structural change that enabled it to nearly double the rate of the "Hindu" GDP growth and as a result, the country was able to sustain this much higher growth rate over 2 decades. Pakistan's economy, on the other hand, has stayed on a roller coaster with periods of high growth followed by periods of sluggish performance. Today, it is entering another period of low growth.

There are a number of reasons why Pakistan was not able to sustain high growth rates. A significant share of the investment that financed growth spurts came from the influx of foreign capital that augmented the low level of domestic savings, most of it from the United States. External finance became available to compensate the country for the strategic help it provided America. The Pakistani government closely aligned the country with America in the 1960s in support of Washington's efforts to deny additional strategic space to European and Asian communism.[3] The country was rewarded for its loyalty with large amounts of military and economic assistance. In the 1980s,[4] Pakistan chose to become the front-line state in the American effort to expel the Soviet Union from Afghanistan. Once again, the reward was military and economic assistance. More recently, Pakistan was recruited to join America's war on terror and for its support was given an estimated $10 billion of assistance over the 6-year period from 2001 to 2007.[5] In other words, the country did little to generate high rates of economic growth by using its own resources. It also did not improve the quality of governance or ensure continuity in policymaking. These factors have been identified by economists as important contributors to growth.

There is now a vast body of literature that addresses the question: What makes economies grow?[6] Apart from the generation of domestic resources to sustain a high level of investment, two other determinants are very important: well-developed human resources, and institutions that can support development. Successive administrations in Pakistan did little to create these two conditions, resulting in an economy that grew only when large amounts of external capital became available. The rate of growth plunged when, for whatever reasons, the quantity of resources being made available declined. Pakistan has not been through the kind of paradigm shift that made it possible for India to climb on to a high growth trajectory.

Even though the economy has continued to be volatile, it did make considerable progress. Its structure changed quite significantly. As shown in Table 3, since Pakistan's establishment as an independent state, the economy, as well as the society, was basically rural. Agriculture was by far the most important sector of the economy, representing nearly 62 percent of the GDP. Manufacturing contributed a very small amount, less than 7 percent. Now, the share of agriculture has declined to below 22 percent, while that of manufacturing has increased to more than 18 percent. The service sector is now the largest part of the economy by far, contributing more than 50 percent of the GDP.

While agriculture still remains a significant source of employment, its share has declined. As shown in Table 4, it employed over 60 percent of the labor force in 1949-50; 6.7 million out of the total work force of 10.3 million. In 2005-06, agriculture's share of employment had fallen to less than 45 percent. The number of people employed in agriculture tripled in 60 years, from 6.7 million to 21.3 million, but in the same period the number of people in nonagricultural employment increased seven-fold, from 3.6 million to 26.3 million. As a result, the structure of the economy is considerably different from the one the country inherited at the time of independence.

Table 3. Sectoral Shares of GDP.

	1949-50	1969-70	2005-06
Agriculture	61.7	38.9	21.6
Mining and Quarrying	0.1	0.5	2.6
Manufacturing	6.9	16.0	18.2
Large Scale	12.5	12.7	
Services	25.5	38.4	52.3
Wholesale-Retail Trade	(9.3)	(13.8)	(19.2)
Finance and Insurance	(0.2)	(1.8)	(4.6)
Public Administration and Defense	(4.7)	6.4	5.8
Construction	(5.8)	4.2	4.3
Electricity and Gas Distribution	—	2.0	3.0

Source: For 1949-50, J. Russel Andrus and Aziz F. Mohammed, Pakistan's Economy, London, UK: Oxford University Press, 1957. For 1969-70 and 2005-06, various issues of Pakistan Economic Survey, Government of Pakistan.

Another change—not as significant as those noted above, but important nevertheless—is the larger role women play in the economy by participating in the work force. As shown in Tables 4 and 5, women were almost totally absent from the work force at the time of Pakistan's birth. In the late 1940s, there were only 300,000 women formally recognized as participants in the labour force, only 3 percent of the total work force. Most women stayed home in this period, one reason why Pakistan, at less than 31 percent of the work force, had one of the lowest worker participation rates in the developing world. That changed over 60

Table 4. Economic Distribution of Total Population.

	1949-50 (Million)			1949-50 (Percent)		
	Male	*Female*	*Total*	*Male*	*Female*	*Total*
Civilian Labour Force	10.0	0.3	10.3	29.7	1.0	30.7
Agriculture	6.5	0.2	6.7	19.4	0.7	20.1
Nonagriculture	3.5	0.1	3.6	10.3	0.3	10.6
Total Population Percent in Labour Force	33.7					

Source: Chap. 7, Pakistan Economic Survey, Islamabad: Government of Pakistan, 2005-07.

years and the participation rate has increased to nearly 45 percent. There was a 20-fold increase in the number of women taking part in the work force. In 1949-50, only 300,000 women were formally part of the work force; 55 years later, their number had increased to over 9.2 million. It should be stressed, however, that the number of women formally recognized to be working is considerably less than those who actually work. In most economies, not just in the developing world, women's work in the house is not recognized as work in a formal sense. In a country such as Pakistan, women put in hard work in both rural and urban areas, particularly among the lower income groups. Even young girls labour hard to help their mothers take care of their younger siblings. Women put in many hours a day caring for animals, which are an important source of income for poor households.

The last significant change I would like to recognize is a large increase in the urban population. In 1947, the proportion of Pakistanis living in urban areas was no more than 12 percent,

Table 5. Economic Distribution of Labour Force, 2005-06.

	2005-06 (Million)			2005-06 (Percent)		
	Male	*Female*	*Total*	*Male*	*Female*	*Total*
Agriculture	14.9	6.4	21.3	38.4	69.9	44.8
Manufacturing	5.1	1.3	6.4	13.4	14.0	13.6
Construction	2.8	0.03	2.8	7.4	0.3	5.9
Whole Sale-Retail Trade	6.6	0.17	6.7	17.3	1.8	14.1
Transport	2.8	0.03	2.8	7.2	0.3	
Community, Social and Personal Services	5.2	1.2	6.4	13.9	13.5	13.5
Other	1.1	0.1	1.2	2.0		
Total Labour Force	38.4	9.2	47.6			
Population	0.7	79.3	160.0			
Percent in Labour Force	47.5	11.6	29.7			

Source: Chap. 7, Pakistan Economic Survey, Islamabad: Government of Pakistan, 2005-07.

some 3.6 million out of a total population of 30 million. The arrival of 8 million refugees from India, 2 million more than the 6 million Hindus and Sikhs who migrated in the other direction, resulted in a significant increase in urban population. In 1951, when the first population census was taken, 17.6 percent of the population lived in urban areas. By 1972 the proportion of the urban population increased to 22.4 percent, with a further increase to 28.3 percent in 1981. The last census taken in 1998 estimated the proportion of people living in the urban areas at 35.4 percent.

The Current Economic Situation; Macroeconomic Imbalances Return[7]

In 2008, Pakistan's economy is once again at a critical juncture. After a period of strong economic expansion, relative macroeconomic stability, and increased foreign investor confidence during the years 2003-06, the country is facing very serious economic strains and a number of social challenges. Macroeconomic indicators deteriorated very sharply over the last few years. Inflation touched record levels in the first 9 months of 2008 following 3 previous years of high single-digit increases in the level of prices. This is despite the fact that the sharp increases in international oil prices during most of 2008 were not fully passed on to consumers and the price of wheat for urban consumers was subsidized. The burden of high prices, especially of basic food items, became intolerable for poor households. One of the primary causes of inflation since 2004 may have been monetary in character, but in 2008 they acquired a structural nature, given the high dependence on imported energy.

Over the same period, poverty levels increased again.[8] There was some decline in the poverty rates from 1999-2005, but the unprecedented rise in food prices since 2004, along with the shortage of wheat flour and a slowing economy, eliminated any gains that had been made. Also, there was evidence that labour absorption was limited despite rapid economic growth in the 2002-07 timeframe.

Structural problems constraining long-term growth came dramatically to the forefront in the first half of 2008 with major power shortages and largescale load shedding. In addition, the erosion of the competitiveness of the country's dominant exports, textiles, and clothing, and a sharp slow down in export growth since 2006-07 led to a large increase in the trade imbalance and limited the prospects for growth in labourintensive manufacturing.

Given this backdrop, I will take stock of the economy by focusing on:

- the immediate financial problems arising out of large and virtually unsustainable twin fiscal and balance of payments deficits;
- a high and rising rate of inflation, especially in food and energy prices;
- a slowing down of the economy, especially in the sector of manufacturing, and the need to remove the principal constraints on long-term growth like the power deficit and water scarcity;
- widespread poverty incidence, as well as growing income disparities, among income groups and across regions; and,
- the governance and institutional problems that not only hamper productivity and growth but also prevent the poor from accessing government resources, public services, and participating in government decisionmaking.

In attempting to assess the present position, this chapter analyzes the short-term causes of the unraveling of the economy in the first half of 2008 as well as the underlying longer-term issues that continue to impede economic growth and social progress. Both perspectives are critical because not only is Pakistan quite a distance away from matching the record of the Asian tigers, Hong Kong, Singapore, South Korea and Taiwan, but also is also seriously falling behind India.

The second and central objective of this chapter is to outline a comprehensive and integrated economic and governance strategy that will facilitate the tackling of the previously mentioned challenges and that will require the urgent attention of the new economic and political leadership.

The chapter recognizes that efforts to restore macroeconomic stability from the position of almost uncontrollable fiscal and balance of payments deficits could dampen short-term growth and investment and make the addressing of poverty and distribution issues harder during the period of adjustment. The agony of a sharp adjustment is unavoidable though it should be possible through public policy measures and well-designed interventions to protect the poor who account for around 10-11 percent of total private consumption. In the absence of a strong adjustment the country runs the risk of a deep financial crisis with catastrophic consequences for its citizens.

In the longer run, the goals of financial stability, rapid growth, and fairer income distribution can be achieved. These objectives are not only consistent with each other but can be mutually reinforcing and interdependent with the appropriate public policies and resilient national institutions.

In looking at future prospects for growth and seeking better distribution outcomes, this chapter highlights both the gains made in the last few years as well as the many un-met challenges and unexplored opportunities, especially in the context of development in the global economy. In looking at the choice of policy instruments to advance the economic and social agenda, the chapter stresses the need to move simultaneously on a number of fronts because of the interlocking and mutually reinforcing effects of many policy and institutional changes. For example, improvements in governance could partly alleviate the pain of economic adjustment.

The Road to the Present Crisis

The macroeconomic situation unraveled very quickly. The fiscal deficit (excluding grants) grew eight-fold over the 4-year period between 2004 and 2008, approaching 8 percent of GDP. The first finance minister of the coalition government that took office in March 2008, in fact, projected the fiscal deficit at 9.5 percent of the GDP on the basis of current trends. However, an adjustment of 1.5 percent of the GDP was made by the government in the second quarter of 2008, primarily because of a rationalization of the Public Sector Development Programme (PSDP).

The current account of balance of payments which had a surplus of almost 2 percent of the GDP as late as 2003-04 reached the record level of $12 billion or 7.5 percent of GDP in 2007-08. Here again, the then finance minister projected a higher deficit at above 9 percent of the GDP. The large deficit cut into the foreign exchange reserves at a most worrying rate. In the first 4 months of 2008, the decline in reserves accounted for nearly 40 percent of the current deficit. By the end of 2007-08, the foreign exchange reserves had dropped to $11 billion and were $5.5 billion below the level at the end of October 2007. By October 2008, the reserves were estimated at only $6 billion.

Pakistan's present predicament is the result of a combination of factors; large exogenous shocks, wrong or the absence of policy responses, and a neglect of emerging structural problems in three key sectors—energy, agriculture, and exports.

The negative shocks, including a devastating earthquake in 2005, the inexorable rise in international oil and food prices, especially of grains and edible oil, have all placed a huge tax on the economy and have effectively reduced the real growth of income in Pakistan by about 2 percent per annum on average during the last 4 years. Another increase in oil prices could cost Pakistan another 2 percent of its GDP in 2009.

The policy response to this state of affairs has been poor or misguided. These developments required a major adjustment in consumption and possibly investment plans. But the need to reduce aggregate demand, especially by reining in the expansionary monetary policy, was ignored partly because the revenues from the privatization of the economic assets owned by the government and sovereign borrowing in world markets were easily available to finance growing deficits and partly because delivering high growth was considered a political imperative for winning the elections of February 2008. It was a false assumption, and the ruling party lost the elections and the burden on the economy remained.

Domestic absorption of resources increased very sharply from 2004-07. Real consumption and investment collectively increased by 35 percent over the 3 years in contrast to the growth in national income, adjusted for terms of trade loss, which only grew by little more than 25 percent. The imbalance was directly reflected in the deterioration of the balance of payments. Consequently, the propensity to import jumped markedly three times during the last 5 years as the economy sucked in more resources from abroad.

The hardest challenge will be to avoid a balance of payments crisis that would further shake the confidence of foreign investors and citizens and could accelerate capital flight as well as limit Pakistan's access to the international capital markets. The rapid accumulation of foreign assets resulting from the quantitative jump in home remittances and the emergence of a current account surplus after September 11, 2001 (9/11) encouraged the government to resort to an expansionary monetary policy from 2002-03 onwards. This policy was too easy for too long and led to a precipitous fall in interest rates which promoted the rapid growth of consumer financing. By 2004-05 there was evidence that the economy was beginning to overheat, as evidenced by the inflation rate jumping to over 9 percent, even in the absence of international inflation and rising commodity prices. Expansionary policies did succeed in reviving growth, but they put the economy on a highly inflationary path. After nearly 4 years of high single-digit inflation, inflationary expectations have become built into the behaviour of economic agents, especially with regard to consumption and savings. Even stronger policy action is now required to counter these expectations.

Superimposed over the history of inflation is the recent upsurge of oil and food prices. This gave rise to upward spiraling prices, even though full domestic adjustments to higher international prices have not yet been made. The inflation of food prices was running at 20 percent in the first half of 2008.

Fiscal policy began to reinforce monetary policy and added to inflationary pressures. On the surface, the actual deficits of 4.3 percent of GDP (including earthquake related spending) in 2005-06 and 2006-07 may not appear excessive. But the way they were financed triggered further strong monetary expansion. The government experienced difficulty since 2005-06 in meeting the growing domestic borrowing amount from the market on longer-term Pakistan Investment Bonds (PIBs) without offering higher interest rates. It thus resorted to the low cost alternative of borrowing huge amounts from the State Bank of Pakistan (SBP), the central bank. This moderated the cost of government borrowing (thus helping to keep interest payments on domestic debt in the budget low), but it also contributed to higher rates of monetary expansion by creating excess liquidity in the banking system.

Despite the measures taken to tighten monetary policy in 2006-07, broader money grew by over 19 percent during the year, even somewhat higher than the average annual rate in the previous 3 years. During 2007-08, the growth of the money supply was running at approximately 7 percent, but this was mainly due to a decline in foreign assets. Government borrowing from SBP during July-March was at the record level of almost 4.5 percent of the GDP.

The SBP correctly tightened monetary policy in early 2008. The space in which the central bank can manoeuvre should be expanded by largely eliminating the sizable amount of government borrowing. Market borrowing by the government through the PIBs will help to identify the true cost of public debt service, improve the interest rate structure, and thus encouraging savings and reducing the supply of reserve capital.

The major instrument of economic adjustment, however, must be fiscal policy. Fortunately, fiscal adjustment can take place in an environment much more favourable than in the 1990s when elected governments had little fiscal space because of the extraordinary burden of interest payments on public debt. Real public noninterest spending, which had shown no increase in the decade of the 1990s because of the growing burden of interest payments, expanded, adjusted for inflation by over 60 percent from 2004-07, and would show a further increase this year because of large subsidies for oil.

The details of a desirable fiscal adjustment are discussed below. A strong fiscal adjustment and a tight monetary policy will send strong signals to the markets that Pakistan seriously intends to tackle the disequilibrium in its foreign transactions and avoid any disruptive change in the value of its currency or a flight of capital.

Growth: Emerging and Structural Constraints

Since independence, Pakistan's average annual growth rate has been less than 5 percent per annum, much below the 8-9 percent growth enjoyed by East Asian countries. Even in boom periods, average growth never exceeded 7 percent per annum. Pakistan has some fundamental demographic governance and growth problems that have kept it from joining the ranks of the Asian Tigers. These problems have deep roots, which include a high population growth rate; a low rate of savings, and consequently inadequate investment not only in human capital but also in infrastructure, industry, and agriculture; a weak industrial and export structure dominated by cotton based exports; an ambivalent attitude towards the private sector and the absence of liberal economic framework till the early 1990s; a level of defense spending that the country could ill afford; inability of the government to collect enough revenues; a major neglect of human development; an inability to develop viable democratic political institutions and effective governance structures resulting in over-centralized decisionmaking, weakening public institutions and rule of law, public corruption, and lack of accountability.

These problems notwithstanding, there are several positive indications that could signal a better economic future for the country. They include changing demographics; liberalization, privatization and reform of the financial system; and increased confidence in the economy, which helped to energize the private sector and increased foreign investment flows for some time, all symbols of increasing economic efficiency. Greater depth in the capital market has enabled it to handle the recent economic crisis well. However, these positive trends will need to be reinforced, something which could have been done when the new government announced its budget proposals for 2008-09 fiscal year. Unfortunately, this did not happen.

There are still major problems that relate to the private sector development and public sector priorities. There is a crisis in the electricity sector. Insufficient investment in generation and distribution and inefficiencies not only increase the costs for the private sector by requiring alternative generating capacity, but also result in large losses for public entities, which are a significant drain on their resources. Government policy favors the traditional private sector industries, such as textiles, far too much. The medium and small industries, though faring better than before, are not getting the support they deserve. Also, large foreign investment flows are taking place in the areas that do not contribute directly to export development. Since export growth remains critical for Pakistan's development, an imbalanced pattern of foreign investment could prove costly in the long run.

The poor in Pakistan continue to face markets, institutions, and local power structures that discriminate against their access to resources and public services and that impede their influence on governance decisions. Due to the unequal access to capital and land and labour markets, inequality and poverty are built into the structure of the growth process itself. On the basis of new estimates, statistics have been provided for the first time on the incidence of poverty from 2005 to 2008, with forward projections for the next 4 years. The evidence shows that after a decline in the poverty rate from 2000 to 2006, poverty levels have since increased neutralizing the earlier gains, as food inflation accelerated and GDP growth declined. For the Musharraf period as a whole (1999-2008), the percentage of population below the poverty line increased from 30 percent in 1998-99 to almost onethird currently, with an additional 16 million people being pushed into poverty during this period. The central policy lesson of the economic performance of the Musharraf regime is that poverty levels increased in spite of high GDP growth in later years because growth was heavily tilted in favour of the rich and high food inflation was not controlled. Recent analyses highlight the importance of controlling food inflation and at the same time bringing about the institutional changes necessary for pro-poor growth.[9]

Placing the Economy on a Sustainable Path

If Pakistan is to get on to a sustainable development path, the government has to follow a different route. The main conclusion of this chapter is that growth, equity, and financial soundness must be pursued simultaneously. Listed below are some strategy changes that the government should take:

- Make radical macroeconomic adjustments by eliminating energy and wheat subsidies, significant cutbacks, and a restructuring of public spending, which has grown sharply during the last 5 years. The government should make a determined effort to generalize tax revenue from the segments of the society whose taxation rates have been drastically cut and those who escape the tax net, while improving incentives for savings and discouraging luxury consumption.
- Substantially expand the safety net for the poor by allocating significant resources, perhaps as much as rupees (Rs.) 50 billion, to minimize the impact of the elimination of the wheat subsidy and potential increases in food prices.
- Make the expansion and diversification of exports a key tenet of any growth revival strategy with a special focus

on agriculture and promising labor-intensive manufactured exports, based on geographical comparative advantage.

- Strengthen devolution by shifting governance and expenditure from the center to provinces and from provinces to local governments.
- Expand education at all levels, especially by improving the quality of public education and increasing the access by relatively poorer families to the privately run educational institutions.
- Increase outlays for research and development, especially agricultural research in recognition of the fact that high growth in Pakistan will require a faster pace of productivity improvements and efficiency gains because low domestic savings remain a major constraint on investment.

The economic and political costs of adjustment in terms of consumption restraint and popular support will be real but should not be exaggerated. The growth of GDP could decline to 5 percent per annum for a year or so but the combination of a necessary reduction in the current account balance of payments of at least 2.5 percent of GDP and a significant cut in current government expenditures, would make moderate increases in real consumption of 0.5–1.0 percent per capita possible for most income groups. Considering that average private consumption per capita grew by well over 20 percent during the period 2003-07, the transition should be manageable, provided the burden of adjustment is equitably distributed.

It needs to be emphasized that if the macroeconomic adjustments are simultaneously combined with measures that improve the fairness of policies, increase participation, and employ the people following the return to democracy, a temporary slow down in consumption growth might be publicly acceptable.

Greater control over a somewhat smaller pie would be welcomed by the lower tiers of government because the pain of expenditure cuts would by balanced by gains in efficiency and a reorientation of priorities towards the poor.

Policy changes necessary to achieve more sustainable and inclusive growth are elaborated below.

Balance of Payments Adjustment

The current account deficit is so large and the need for curtailing it, as well as curbing speculative pressures on the exchange rate, is so urgent that fiscal and monetary policies would have to be strongly supported by trade, exchange rate, and foreign exchange reserve policies and confidence-building measures such as adopting a strong export orientation and clearly articulated external finance strategy.

As mentioned above, the current account balance of payments deficit in 2007-08 was around 7.5 percent of GDP. This should be reduced to 5 percent of GDP in 2008-09 and 4 percent in 2009-10. Pakistan can safely run an account balance of payments deficits of this latter magnitude provided export growth recovers to at least 10 percent per annum, private transfers remain strong, and the supply of concessionary assistance ample. Equally important would be to limit the deficit to 4 percent of GDP and bring the saving-investment gap (a measure of self reliance) to the 15-20 percent range from a record 33 percent imbalance likely recorded in 2008.

The biggest contribution to reducing the saving-investment gap would be the early elimination of negative savings on the general government revenue account, which reemerged and became very sizable (3.5 percent of GDP) during 2007-08. Strengthening incentives for small savers by improving what are now negative returns on bank deposits and improving returns on government saving schemes should also help to curb consumption.

Some restraints on imported consumer goods, especially luxury goods should also be considered through imposition of moderately higher tariffs, as was done in July 2008. Similarly, in reviewing defence expenditures, the postponement of foreign exchange intensive expenditures on weapon systems should be seriously considered. One proposal that merits serious consideration is the levy of a temporary regulatory import duty on all imports, excluding essential imports like food (wheat and edible oil); fertilizer; and petroleum, oil, and lubricant (POL) products, with a higher rate on luxury goods.

The biggest challenge for short-term balance of payments management is to maintain and restore foreign exchange reserves to a level of around $15 billion over the next few months while financing the substantial uncovered gap in financing. More adequate reserves are necessary to ward off the speculators in the liberal global framework in which Pakistan is operating.

With the recent downward trend in the value of the rupee (in the first 9 months of 2008, the rupee-dollar exchange rate fell by 25 percent to nearly 80 to a dollar), Pakistan's exchange rate does not need any significant once-and-for-all realignment. However, it is important to annunciate the policy that the real effective exchange rate will not be allowed to appreciate in the near future. In other words, the much higher rate of inflation in Pakistan compared to its competitors will be allowed to be reflected in the change in the nominal rate against a variety of currencies. Otherwise, the competitiveness of the country's exports would suffer, and import growth will be artificially stimulated. The approximately 6 percent appreciation of the rupee between 2004-05 and 2006-07 may be one factor explaining the slowdown in exports and continued rapid growth of imports.

Pakistan cannot hope to solve its fundamental growth and balance of payments problems without making export development a centerpiece of its development strategy. Rapid export development helps to create jobs, raise wages, and meet the rising obligations of debt servicing and investment income payments.

The major elements in an export-focused strategy should be:

• Strong national commitment at the highest political level.

• Recognition that while textiles and clothing will remain a vital and expanding export sector, it cannot be the future engine of growth. The limits of government support for textiles have been reached, and the industry must learn to be competitive through investments in physical capital and skills.

• Diversification deserves the highest priority, and manufactured goods other than textiles, clothing, and agricultural exports should lead the way and get the necessary government attention and support.

• The role of the state can be crucial in the early stages of export diversification through aggressive targeting of markets and products, improving access, and speedily removing obstacles to trade.

• Foreign direct investment should be especially encouraged in export fields.

Pakistan also needs an external finance strategy and a framework for balance of payment management to complement the Fiscal Responsibility Law passed by the National Assembly in 2006 that put limits on public debt, fiscal deficits, and contingency liabilities. To avoid future balance of payments difficulties, the adoption of a few specific guidelines to implement a viable external finance strategy should be attempted. The first guideline should establish a ceiling of 20 percent of total investment to be financed from foreign savings. A second guideline should place limits on total external debt and foreign investment obligations in relation to total foreign exchange earnings at the present level of 100-110 percent. Another guideline should define the balance between equity and debt financing at 2:1 to meet a given balance of payments gap.

Fiscal Adjustment

The objectives of fiscal policy must be to, first, stabilize the economy by reducing the size of the fiscal deficit and financing it to the extent possible by noninflationary sources. The resulting restraint on aggregate demand can also exercise a favourable impact on the external balance of payments. Second, fiscal policy must play a strong redistributive role and help in reducing the income disparities that have emerged between the rich and the poor and among various regions of the country in recent years. Third, the goal of fiscal policy should also be to sustain the rate of economic growth as much as possible. This would be achieved by generating resources for development and guiding the allocation of these resources towards agriculture and labour-intensive manufacturing with export potential and away from capital-intensive nontradable services in particular.

The task of fiscal adjustment will require drastic changes in revenue and expenditures if the deficit is to be brought down to the sustainable level of about 4 percent of the GDP. This would help avoid an increase in the public debt-to-GDP ratio and eliminate of any deficit on the revenue account, such borrowing should only be used to finance development projects. Fiscal deficit reduction from 8 percent of the GDP to 4 percent should be completed within 2 years if inflationary pressures are to be contained and there is to be less pressure on the external current account deficit. In 2008-09, the target financial deficit must be brought down to 6 percent of the GDP and in 2009-10 to 4 percent of the GDP, with development expenditure sustained at the minimum level of 4 percent of the GDP each year. This would imply a revenue deficit of about 2 percent of the GDP in 2008-09 which will be eliminated in 2009-10, allowing the economy to get back to a fiscally sustainable path consistent with the Fiscal Responsibility Debt Limitation (FRDL) Act of 2005.

Beyond the concern with the size of the fiscal deficit is the issue of how the deficit is financed, especially with regards to the impact of inflation. During the next 2 years, in the period of fiscal adjustment, the government will have to operate strictly within the safe limits of deficit financing. Earlier studies reveal that the scope for "seignorage" in the Pakistan economy is about 1 to 1

1/2 percent of the GDP,[10] if a low single-digit rate of inflation is to be achieved. Other noninflationary sources of financing will have to be used. Up to 1 percent of the GDP can be mobilized from commercial banks through the market flotation of the PIBs of varying maturities. At this level, there should not be a significant crowding out of credit to the private sector.

Beyond this, the biggest increase in borrowing will have to come from nonbank sources; the national savings schemes. In the face of large reductions in the rate of return on certificates, the net inflow has plummeted to only about half a percent of the GDP in recent years. This will have to be raised substantially to between 1 to 1 1/2 percent of GDP by linking the return to that of PIBs, with the expectation that the return will rise by 2 to 3 percentage points. In addition, an effort must be made to develop a secondary market for the PIBs. The offering of positive real rates of return on savings instruments should help in raising the rate of domestic savings and reducing the dependence on foreign savings. The residual deficit will have to be ameliorated by the continued resort to concessionary external assistance at the more or less unchanged level of about 2 percent of the GDP.

The government's strategy should be focused on eliminating the revenue deficit in the next 2 years, while keeping the PSDP at about 4 percent of the GDP to avoid jeopardizing growth. A balanced and politically acceptable strategy will require the same effort to be directed at the containment of current expenditure and mobilization of resources. If the focus is only on the expenditure side, then this will severely limit the possibility of providing social protection to the poor, especially through an expanded programme of food subsidies. On the other hand, if the deficit reduction strategy relies solely on additional taxation, then this could have adverse effects on investment and growth. Consequently, a balance is required.

As highlighted above, noninterest current expenditure has risen rapidly since 1999-2000 by almost 2 percent of the GDP.

The fiscal space that was available earlier has been largely taken up by rapidly increasing outlays on general administration, growing subsidies (especially to the power utilities), rising defense expenditure, and buoyant expenditure on services (particularly by the provincial governments). The bloated size of federal and provincial cabinets during the Musharraf period became symbols of extravagance by the government. The hiring of large number of consultants and retired officials on lucrative salaries with perks, the removal of recruitment bans, the import of large fleets of luxury vehicles, and the expensive foreign missions of dignitaries all became signs of systematic government excess.

As a result, there is significant scope for reductions in current expenditure without adversely affecting the delivery of services. The new prime minister announced a reduction in the costs of running his secretariat by 40 percent after taking office. This example of reduction in nonsalary expenditures should guide all federal and provincial governments and all semi-autonomous organizations and attached departments over the next 2 years. This could yield up to 1 percent of the GDP or about Rs. 100 billion by the end of 2009-10.

The large oil subsidy was to be reduced gradually by the end of 2008, to yield a saving of about Rs. 100 billion. This is essential if demand for POL products is to be contained to maintain the oil import bill at a sustainable level. Of course, the impact on the poor can be limited by a lower increase in the price of products like kerosene oil, high speed diesel oil, and light diesel oil, and a greater increase in the price of gasoline, which is consumed mostly by upper income groups. If the oil price remains at about $100 per barrel, then further adjustments in domestic prices are inevitable if the fiscal and current account deficits are to be contained. This will also release resources for supporting food programmes for the poor and bolster social safety nets.

The big disappointment in the area of public finances is that 4 years of continuously high growth did not lead to a rise in the tax-to-GDP ratio in the economy, which remained stagnant at between 10 to 11 percent. This is despite the buoyancy of major tax bases, like value added in large-scale manufacturing and imports. The explanation for the failure of the tax-to-GDP ratio to rise lies in the decline in effective tax rates. Import tariffs have been brought down to a maximum of 25 percent. Concomitantly, this has also affected revenues from the sales tax on imports. Excise duties have been replaced by sales tax in a number of sectors, and the specific rates have not been fully indexed to inflation.

The large decline in tax rates is from direct taxes. The maximum personal income tax rate was reduced from 30 percent to 20 percent for salaried tax payers and from 35 percent to 25 percent for the selfemployed. Simultaneously, the corporate tax rate has been reduced from 45 percent to 35 percent for private companies and from 50 percent to 35 percent in the banking sector, at a time of sharply rising profitability.

Major tax concessions and exemptions have been granted since 2000 starting with the abolition of the wealth tax. The most dramatic example is the continued tax exemption for capital gains at a time when massive unearned incomes were accruing to the relatively welloff due to the exceptional performance of the stock market in 2005-08 and rising property values. By the government's own estimate, as much as Rs. 112 billion in revenue were lost in 2006-07, almost 1.2 percent of GDP. The cost of other exemptions or concessions adds up to another Rs. 200 billion. This includes the cost of exemptions from import duty, income tax holiday and accelerated depreciation allowance, lack of coverage of sales tax on wholesale and retail trade, effective exemption of a large number of services from General Sales Tax (GST), and the effective zero rating of domestic sales of export-oriented sectors like textiles. If all these concessions and exemptions are accounted for, then the aggregate loss of revenue is roughly Rs. 300 billion.

This is equivalent to over 3 percent of the GDP and about one-third of the revenue actually collected.

The provincial governments have also demonstrated little fiscal effort. Currently, provincial tax revenues aggregate to only half a percent of the GDP. The agricultural income tax, which was introduced in late 1996, has been languishing as a source of revenue despite the rising incomes of large farmers. Consequently, land taxes represent less than 1 percent of agricultural incomes in the economy. The urban immovable property tax also remains underdeveloped, currently exploiting only one-fourth of its revenue potential. Despite the boom in real estate values, stamp duty revenues remained stagnant during the last 3 years, and a capital gains tax on property was not introduced.

The elite has had unprecedented control of the state and granted itself large tax breaks during the last 8 years.[11] There is no doubt that considerable slack exists in the tax system not only for significantly raising the tax-to-GDP ratio, but also for simultaneously achieving a measure of redistribution through the tax system to arrest the rising inequality between the rich and the poor in the country.

A recent study by the Institute of Public Policy identified a series of taxation proposals for implementation by either the federal or provincial governments over the next 2 years with a potential yield of up to 2 percent of the GDP by 2009-10.[12] These include an excess profits tax, higher tax on private companies; introduction of a capital gains tax; a more progressive personal income tax; higher taxation on imports, especially luxury goods; a broad-based services tax; and development of provincial taxes.

Overall, the proposals outlined above are oriented toward mobilization of revenues from direct taxes or from indirect taxes on goods and services consumed by upper income groups.

Implementation of these proposals will make the tax system more progressive while improving public perception about a more equitable distribution of the tax burden.

Promoting Inclusive Growth

At this stage it would be useful to consider some structural weaknesses in the growth process and indicate how they could be overcome. The purpose is to understand how high rates of growth could be attained and sustained over time, while ensuring that the benefits are spread more widely.

Sectoral Strategy

Sectors like banking, telecommunications, and automobiles, which were in the lead during the last 5 years, will not keep the economy on a high growth track for very long. They will also not do enough for the poor. The number of jobs created by these sectors and the types of employment they generated did little to reduce the incidence of poverty, as demonstrated earlier. In addition, the pattern of growth widened interpersonal, interprovincial, and intraprovincial income disparities. Increases in such disparities usually lay the groundwork for social and political instability, a development Pakistan does not need at such a difficult period in its history.

An increased focus on the basic commodity—producing sectors of the economy—agriculture and manufacturing—is needed. This change in sectoral focus will require actions from all three tiers of government—federal, provincial, and the local—as well as from the private sector. This raises the question: how could this be done?

Agriculture

Pakistan has one of the best endowed agricultural sectors in the world. It has one of the world's largest contiguous irrigated areas; it has rich soil created by deposits made by rivers over thousands of years; it has hard working farmers who have shown their ability to absorb new technologies when presented with the opportunities to do so; and it now has rapidly growing internal and external markets for the products produced by high value added agriculture. While the agricultural system is entirely operated by the private sector, these operators are responsive to the incentives provided by the public sector. The public sector, therefore, has an important role to play. In this context, three aspects of public policy are particularly important.

Among the more important ones are the price signals embedded in public policy. These have a profound impact on cropping patterns. The most important price signal the government provides is the wheat procurement price. Wheat is the country's most important crop. The anticipated income that farmers receive from cultivating wheat significantly affects what else they grow. The federal government should continue to handle the procurement price of wheat while monitoring the level and expected trends of international prices. The recent rise in world wheat prices represents a trend caused by the increase in demand for food grains in rapidly growing populations such as China and India and the increasing return given for bio-fuel production by such large consumers of energy as the United States. The rise in the price of wheat has affected the prices of other food grains—commodity prices normally move in tandem—and has changed the sectoral terms of trade in favour of agriculture. The benefit of these should be passed on, to the maximum extent possible, to agricultural producers.[13] For that to happen, there should not be a large difference between the government's procurement price and the price in international markets. In the context of the need to make fiscal adjustment, an increase in the price of wheat will have to be mitigated by directly helping the poor through initiatives such as the Baitul Maal (a Pakistani nongovernmental organization) and Food for Work programmes.

The next important area for government promotion of agriculture is in improving the technological base. Here, Pakistan seriously has lagged behind. Very little research and development work gets carried out by the private sector, not surprising given the absence of large commercial operators. The little research that gets done is by the public sector, but it is too widely scattered among too many government departments and agencies to be effective and does not reach the farmers. The result is that Pakistan has developed gaps between average yields and yields obtained by the best farmers; between the best farmers and those obtained by research institutions; and between research institutions and those obtained by farmers in the large agricultural systems in other parts of the world.[14] The role of government must help to close the technology gaps. This can be done in two ways: (1) by focusing on the development of research in agricultural universities (an approach followed by the United States) and (2) by establishing crop or product specific research institutions (as is being done by China). At the same time, incentives should be provided to the private sector to encourage research and development.

The third role of the state in promoting agriculture development is to provide the infrastructure the sector requires. Pakistan has inherited an elaborate irrigation system, and impressive improvements to this network were made as a part of the agreement with India on the distribution of the waters of the Indus River systems. But these were replacement works; they did not result in bringing much additional land under cultivation.[15] However, not enough attention was paid to maintaining this system and for improving it to preserve water. In recent years, the Punjab and Sindh governments, encouraged by the World Bank, have begun to devote sizeable resources to maintenance.

Punjab, in particular, has gone further by producing a fairly elaborate system of information available on the internet that can be used to monitor the flow of water. This information is available to both users of water as well as those who manage the system. As the provinces strengthen their capacity to get engaged in economic development, it is important that irrigation system maintenance and efficiency improvements are high government priorities. The resources being committed to it by the public sector should be protected during the period of adjustment as discussed above.

Livestock husbandry has become an increasingly important part of the agricultural sector, and the modernization of livestock markets need to be promoted. The sector contributes almost 50 percent of agriculture's gross output, which translates into a contribution of over 10 percent of the GDP. It engages 35 million people in the rural economy and provides almost 40 percent of the total income of the farming community. The sector is dominated by small operators; those owning less than two animals account for slightly more than two-fifths of the total population of cattle and buffaloes. As in the case with the crop sector, yields are low. The government estimates the yield gap—outputs of the current livestock population compared with the output obtained in more developed systems—at between 60 to 80 percent. The reason for low productivity has been identified as inadequate and poor quality feed and fodder, limited animal health coverage, widespread breeding of genetically inferior livestock, poor marketing infrastructure, shortage of trained manpower, inadequate incentives for small producers, and a lack of extension services.

Improving yields in the livestock sector would make a significant contribution to increasing value added in agriculture. It would also have the profound impact of reducing the incidence of poverty in the countryside. A strategy aimed at achieving this objective should provide better education and training to the people engaged in work with livestock and better health coverage for animals. For the quality of food and fodder to be improved, the flow of credit to livestock owners also needs to be increased. At this time, 90 percent of bank lending to agriculture goes to the crop sector, with the livestock sector receiving 10 percent. The proportion going to the latter needs to be raised to better reflect its value added.

Manufacturing

The other objective of the strategy for developing the real sectors of the economy is to encourage the growth, modernization of, and exports from small and medium-sized enterprises. Numbering some 3.2 million, these enterprises follow a long tradition of entrepreneurship and craftsmanship, particularly in the provinces of Punjab and the Northwest Frontier Province (NWFP). The sector represents almost 30 percent of the manufacturing output, over 5 percent of GDP and 20 percent of nonfarm rural employment. An industrial policy aimed at the development of this sector would also have three components.

The first is the identification of subsectors and enterprises within these subsectors that will receive government assistance. Not only should the chosen enterprises receive subsidies, but they should also be exposed to the opportunities available in the rapidly evolving global systems of production and trade. The second is the facilitation and development of the chosen sectors. The third is to help the chosen sectors with financial support. While the second component of the strategy is in the mandate of the Small and Medium Enterprises Development Authority (SMEDA), a federal corporation established in 1998 to promote the development of the long neglected industries—the first and the third components have not engaged the state. This needs to be remedied.

Underscoring the need to make additional financing available to the SME sector, it is important to note that the government should not provide the subsidized credit, neither should it direct the banking system to finance these enterprises. What is needed is the introduction of relatively new instruments of finance into the sector. These include private equity and venture capital funds that share risks with the owner-entrepreneurs in which they invest while expecting high returns for themselves. Making these finance instruments available to the SME sector would help to liberate the generations of untapped capital potential in small enterprises, while examining the country's underdeveloped capital markets.

Human Resource Development

The priority areas to be addressed by public policy change will only produce the desired results if the quality of the human resources available in the economy is improved. Concentrating on developing human resources would mean placing focus on at least four areas of public policy. These are improvements in primary and secondary education, increasing literacy rates for women, providing modern skills to a large proportion of the country's youth, and creating synergies between the research and development of the various sectors of the economy. After years of experience with using human resource development as an important contributor to growth, practitioners have realized that they need to promote not only universal primary education but also getting children to stay in school for at least 8 to 10 years. It is only then that children are prepared to enter institutions of higher learning or to make a contribution to the economy by entering the work force. Past emphasis has mostly been on primary education. Such was the case in the World Bank-sponsored Social Action Program (SAP), which was implemented in Pakistan in the early 1990s. While the programme increased the rate of enrollment in primary schools, it had a negligible impact on improving the country's human resources. The failure of SAP to achieve its promised results was because of institutional failure at the level of weak education departments in the provinces, which were unable to efficiently absorb the resources that were made available to them. Development experts have reached the conclusion that pupils need much more than 5 years of schooling to change behaviour and to prepare themselves for the modern sectors of the economy. As a result, priorities must be shifted more towards secondary education.

The provinces will have to play a key role in promoting agricultural development and SMEs. For this, they will need more authority. With greater economic authority, the provincial governments will be in the position to lend strategic coherence to their development programmes. Their focus should be oriented towards building analytical and planning capacity, establishing a stronger relationship with the private sector, and emphasizing opportunities that could emerge from favourable international developments.

Poverty Programmes

While the implementation of an inclusive growth strategy of the type described above will strengthen the process of poverty reduction in the medium term, there is need to ensure that the aggregate demand management of the economy and the withdrawal of subsidies does not lead to a sharp rise in poverty.

Strong social safety nets will have to be put in place to ensure that there is adjustment with a human face. In particular, food security for the poor will have to be protected to avoid a reduction in nutrition levels. This can best be achieved by a combination of cash transfers and employment guarantees. Hitherto, the subsidized sale of food items through the utility stores has been fraught with problems of limited coverage, especially in the rural areas, and ineffective targeting.

The cash transfer scheme will primarily benefit more vulnerable groups such as the disabled, the elderly, female-headed households, and widows. The employment guarantee programme can provide an opportunity for able-bodied workers to earn an income, especially in the off-peak season.

In his 100-day plan, the new Prime Minister announced the intention of his government to launch an employment guarantee programme in the underdeveloped districts of the country. Baitul Mal already runs a cash supplement scheme for food support, which can be scaled up to cover a larger proportion of poor households.

In 2006, India was the first country in the world to introduce a national rural employment guarantee programme, based on the experience gained in one state, Maharashtra. The programme is being run in over 40 percent of the districts with the help of Panchayati Raj (local communal assemblies) institutions. It is estimated that at full coverage, the programme could cost up to 2 percent of India's GDP. A similar programme should be tested in a few of Pakistan's poorest districts. It would also be appropriate to give this programme the characteristics of a food for work initiative, so that the poor workers are automatically protected against inflation.

An ideal cash transfer programme should be based on the identification of the poor beneficiaries by the lowest tier of local governments, the Union Councils. Efforts must be made to reduce the transaction costs and programme inefficiencies. Initially, Baitul Mal would be given funds to at least double the coverage of the programme to about three million households, with cash support per household of about Rs. 1,000 per month. It is expected that the total cost of running the two programmes for protecting the poor will be in the vicinity of Rs. 50 billion, and implementation of these programmes must proceed on a priority basis.

Over and above the sectoral strategies and strategies aimed at alleviating poverty, the government, as discussed above, will also have to use other policy instruments to encourage and promote growth. One such instrument is fiscal policy. It is vital that in the process of fiscal adjustment that the level of development expenditure does not fall sharply as happened in the earlier years of this decade. Not only will the size of PSDP have to be sustained at a minimum of 4 percent of GDP until 2009-10, but there will also have to be a more strategic and rational allocation of development funds to projects. Clearly, public investment in the water and agricultural sectors and in power generation will have to receive higher priority, along with larger allocations for the development of the poorer and more isolated areas of the country. There will be a need for a moratorium on new projects except in the priority sectors.

In addition, fiscal policy will have to be selectively used to incentivize the agricultural and manufacturing sectors as follows:

a. The general sales tax introduced on fertilizer and pesticides needs to be withdrawn so as to improve the ratio of output to input prices and thereby stimulate agricultural production.

b. Power load shedding has adversely and significantly impacted production, especially in the industrial sector. As such, a tax credit (chargeable against all tax liabilities) should be made available to manufacturing enterprises on the capital cost of captive power generation or energy-saving equipment. A similar tax credit can be offered on investments in renewable energy.

c. In order to stimulate nontraditional exports, the presumptive income tax on such exports should be withdrawn and the research and development allowance be made available to all exports.

It is expected that these measures will not cost the exchequer more than a quarter of a percent of the GDP or Rs. 25 billion, but could play a significant role in raising production and export rates of the commodity producing sectors and reduce the energy deficit in the economy.

Reducing Regional Disparities

Fiscal federalism will play a key role in addressing the issue of regional disparities. There is the need to ensure that the pattern of intergovernmental fiscal relations evolves in such a way that recognizes the need for more support to the more underdeveloped provinces. There is a constitutional requirement that the National Finance Commission (NFC), representing the federal government and the four provinces, be established every 5 years and should issue an award to resolve two problems. It must first address the vertical imbalance in resources between the federal government and the four provincial governments combined and then secondly, the horizontal imbalance among the provincial governments.

Over the last 7 years since 2002, the NFC has failed to arrive at a consensus on a new award to replace the one given in 1997.[16] Consequently, President Musharraf promulgated an interim arrangement for transfers that came into effect in 2006-07. With respect to the 1997 award, there are two significant changes. First, the share of revenues provided to the provinces from the divisible pool of revenues has been increased from 37.5 percent to 41.5 percent in 2006-07, rising to 46.25 percent by 2010-11, and second, these benefits have now been extended to all four provinces on the basis of predetermined shares, whereas in 1997 they were given to NWFP and Balochistan. Overall, it is expected that revenue transfers from the divisible pool and grants-in-aid will constitute 50 percent of the revenues in the divisible pool by 2010-11. The sharing of revenues in the divisible pool on the basis of population and the coverage of straight transfers remains unchanged.[17]

The basic issue is whether over the last 7 years fiscal transfers have been adequate and if the goal of fiscal equalization has been achieved, and the two smaller and less developed provinces, NWFP and Balochistan, have received higher transfers on a per capita basis. Incidentally, in the Pakistani context, straight transfers have historically been performing an equalization function. The NWFP has access to hydroelectricity profits and Balochistan has revenue from natural gas, which raise per capita transfers significantly.

A review of the four provincial budgets reveals that transfers have probably been high enough to support an increase in their combined share of public expenditure. But a more in-depth analysis reveals that provincial expenditures have risen because of greater resort to borrowing, which is now financing as much as two-thirds of development expenditure. Also, the share of total transfers to provincial governments in federal revenues (tax plus nontax) has remained virtually unchanged at 35 percent over the last 7 years.

What has been happening to fiscal equalization? The overall growth in per capita transfers of all types to the provinces from 2000-01 to 2006-07 has been 144 percent for Sindh, 106 percent for NWFP, 103 percent for Punjab, and 75 percent for Balochistan. It appears that the process of fiscal equalization has largely broken down with the highest growth in transfers going to the most developed province, Sindh, and the lowest growth in transfers going to the least developed province, Balochistan. Today, the level of transfers per capita to Sindh is higher than to NWFP, while Balochistan is unable to meet even its current expenditure obligations.

Over the last 7 years, a review of the process of intergovernmental relations reveals the emergence of serious imbalances. This has been one factor contributing to faster growth of the economies of Sindh and Punjab as compared to Balochistan and NWFP. Clearly, there are justifiable reasons why the smaller provinces are dissatisfied with the workings of the federation during the tenure of the last government.

Now that elected coalition governments are in place in Islamabad, and at least three provincial capitals are led by the Pakistan People's Party (PPP), there is urgent need to arrive at an early consensus award that ensures the following:

1. Further expansion in transfers from the divisible pool to cover the emerging sizeable deficits of the provinces with the understanding that they will henceforth face tighter budget constraints with only limited access to borrowings. Provision will also have to be made for higher transfers to cover the costs of taking on more functions by the provinces, as required by the constitution.
2. Adoption of multiple criteria for the determination of transfers from the divisible pool to ensure more fiscal equalization. The collection criteria could also be given some, albeit small, weight. Punjab should be willing to support this plan since the collection rate from the province of apportionable taxes (all taxes, excluding taxes on imports) has approached its population share.
3. Higher grants-in-aid to be made to the more underdeveloped provinces (NWFP and Balochistan).
4. Review of the formula for determination of hydroelectricity profits to NWFP, a long standing demand of the province.

There is no doubt that the transition from an ad hoc award by the President to a consensus-based NFC award will be a major step forward in strengthening the federation and be a key indicator of the success of the newly elected governments.

Decentralizing Governance

One of the more important elements of the strategy developed in this chapter is to give greater operational space to the provinces and to the institutions of local government. The new Prime Minister Yousaf Raza Gillani made an encouraging start in his initial speech before the newly elected parliament, saying that his administration will, within one year, transfer all the subjects listed in the constitution's concurrent list. A great deal of work at the two upper tiers of government will be required for this to be effectively and efficiently done within the current governance structure. This effort should lead to a reduction in consolidated current expenditure by eliminating the duplication of coverage by the government at the federal and provincial levels. In building their own capacity to handle the transferred subjects, the provincial government should place emphasis on the quality rather than the quantity of the staff they employ.

While the decentralization of a significant amount of economic responsibility from the federal government to the provinces would be an important part of the strategy for promoting inclusive growth, it is equally important to continue with the process of devolution to the institutions of local government. A new system established in 2001 is in place and should be continued. However, Pakistan has not been able to develop a viable system of local government because of the continuous experimentation that it has undergone since its independence. Five different systems have been tried in the past. Now, the need is to further develop the existing system rather than create something new from scratch. That said, a number of reforms are needed. The new Prime Minister said that his government will reform the local government system after evaluating how it has performed since 2001. The areas where the structure needs to evolve include the direct election of the *nazims* (indirectly elected managers in the system) to make them more accountable to the people. Governance related services such as law and order should be decentralized, perhaps initially on a trial basis, with responsibility resting with the *nazims*. District service cadres should also be established.

Budget 2008-09: A Lost Opportunity

This is a good time for Pakistan's new rulers to make some decisions that will not only heal the economy but also change some of its structures. Policymakers respond in two different ways to serious economic crises. Those who are bold use the opportunity to deal with the causes behind the crises since most of the time crises are produced by structural flaws in the economic system. They correctly assume that it is best to identify the flaws and remove them from the system and prevent problems from recurring. Those who are less bold implement temporary measures and hope that the underlying problems will not reappear.

Pakistan's policymakers have usually opted for the second approach, preferring short-term fixes rather than deep structural changes. Not surprisingly, the result was a recurrence of crises produced by the same fault lines in the economy. Of the many structural problems faced by Pakistan for the past 60 years, two have been particularly important. The first is poor human development; the second is a low domestic savings rate that did not yield enough resources for the economy to invest. If the economy is to grow at 7 to 8 percent a year—a rate of growth sustained by a number of economies in Asia—it must invest close to 30 percent of the GDP. Pakistan has a domestic savings rate of only 22 percent, which can only support a growth rate of less than 6 percent, perhaps no more than 5.5 percent a year.

The country has done well when domestic resources were augmented by foreign capital flows. The reliance on external savings is not a wise policy to follow since foreign investment is unreliable. On a number of occasions, foreigners have reduced the amount of money they were providing the country. Each time that happened, the amount invested and hence the rate of GDP growth declined. If Pakistan is to stop the volatility of GDP growth, it must increase the amount of resources generated from within the economy.

Domestic savings come in three forms—savings by the government, those by the corporate sector, and those by individual households. Public policy influences all three, especially government savings (or dissavings), and this is where budgets become particularly important.

Ever since the state stopped playing a dominant role in the economy—as happened during the period of President Pervez Musharraf—the available policy instruments used to affect the economy have been reduced to basically two, the fiscal and monetary policies. Whereas the monetary policy is controlled by the SBP, the country's central bank, the fiscal policy is the responsibility of the Ministry of Finance. The SBP can change the monetary stance at any time; the Ministry of Finance usually alters the fiscal policy only once a year when it announces the budget for the year that follows. This is one reason why the budget receives so much public attention. (See Table 6.)

The budget for the financial year 2008-09, announced by the Finance Minister on June 11, 2008, is a particularly important policy statement for two additional reasons. First, it is the first major policy statement that assumed power following the elections of February 2008. Second, it comes at an exceptionally difficult time for the economy. After a 6-year period of relative calm, the economy has become sluggish and unbalanced. This is evident on three fronts—inflation, fiscal deficit, and balance of payments deficit. The budget can influence all of them. Before examining the policy embedded in this budget, it would be useful to look at the way the fiscal deficit has evolved over the last several years.

Fiscal deficit is the difference between government's revenue and expenditure. Government's revenues come in two forms, tax and nontax. Expenditures also come in two forms, current and development. The data provided in Table 6 show some interesting trends over the last decade. The government of President Pervez

Table 6. *Government Revenues and Expenditures (Percent of GDP)*

	1997-98	1998-99	1999-2000	2000-01	2001-02	2002-03	2003-04	2004-05	2005-06	2006-07	2007-08
Total											
Revenue	16.0	15.9	13.4	13.1	14.0	14.8	14.3	13.8	14.2	13.4	14.0
Tax	13.2	13.3	10.6	10.5	10.7	11.4	11.0	10.1	10.6	10.5	11.0
Nontax	2.8	2.7	2.8	2.6	3.3	3.4	3.3	3.7	3.6	2.8	3.0
Expenditure	23.7	22.0	18.8	17.4	18.3	18.5	16.7	17.2	18.5	17.6	19.0
Current	19.8	19.6	18.4	15.3	15.7	16.2	13.5	13.5	13.6	12.7	16.0
Development	3.9	3.3	2.5	2.1	2.8	2.2	3.1	3.9	4.8	4.9	3.0
Deficit	7.7	6.1	5.4	4.3	4.3	3.7	2.4	3.3	4.3	4.3	5.0

Source: Pakistan Economic Survey 2006-07 and my estimates using the budget proposals announced on June 11, 2008.

Musharraf inherited a difficult fiscal situation in 1999. In the 2 previous years, the fiscal deficit had averaged 7 percent of the GDP; 2 percentage points higher than what experts consider to be the sustainable level for Pakistan. While government revenues were reasonably high—about 16 percent of GDP—expenditures were even higher. The difference was in the order of 7 percent mostly because of current expenditures. Nondevelopment expenditure was close to 20 percent of GDP. It was clear to the new policymakers who took office after General Pervez Musharraf intervened that major adjustments had to be made to restore economic balance. They went to the IMF for assistance and received the advice that significant adjustments had to be made.

This was done over a 3-year period by putting the lid on current expenditures, which declined to an average of 16 percent a year, a reduction of nearly 4 percentage points compared to the levels reached in the late 1990s. The most significant reductions were obtained by constraining government employment and putting a cap on government salaries. Further reductions in nondevelopment expenditures became possible after 9/11 when, led by the United States, the donor community reduced the country's outstanding debt by a significant amount. This lowered the interest payments Islamabad paid to its creditors. These adjustments, while reducing the fiscal deficit, also made it possible for the Musharraf administration to increase development expenditure. In the last 3 years of the Musharraf period, development expenditure increased to an average 4.5 percent of GDP. This was 1 percentage point higher than in the late 1990s and almost double the share in the first 3 years of the Musharraf period.

The policymakers faced the same type of challenges while preparing the budget for 2008-09 that confronted the Musharraf government in 1999. The fiscal deficit and the balance of payments deficit in terms of the proportion of GDP had reached unsustainable levels. Adjustments needed to be made to reduce the fiscal deficit by raising taxes and constraining government expenditure. Islamabad also had to deal with the pressure on lower income groups as a result of the increase in the prices of food and fuel. However, in preparing their proposals and presenting them to the National Assembly, the policymakers opted for the approach adopted by their predecessors: they did not attempt to bring about structural changes in the economy, preferring to tinker at the margin. That said, there were some attractive features in the budget.

An attempt was made to help the poor by creating a new fund to provide them with cash transfers. On the revenue generation side, some rates were adjusted to increase the burden on the rich. Some luxury items will cost more, and the higher income groups will pay more for some of the services they use, such as cash withdrawals from the banking system. These changes will help to raise some additional tax revenues for the government. I have calculated the impact of these proposals on government revenues and expenditures and the size of the fiscal deficit. These calculations are shown in Table 6. They are more reasonable than the estimates provided in the budget.

The budget proposals may also reduce conspicuous consumption by the rich. The changes, however, will be marginal and will not make much of a difference to one of the most important structural weaknesses in the economy: dependence on external flows for a significant proportion of gross investment.

What could the policymakers have done to deal with the structural problems that continue to affect economic performance? Those who made the budget could have taken four additional measures. One, they could have created fiscal space for the provinces, thus creating the opportunity to both raise additional government revenues and grant greater provincial control over public expenditure. This could have laid the basis for increasing domestic resources by bringing government closer to the people. Two, they could have significantly reduced current government expenditure by eliminating some of the functions it should devolve to the provinces. Three, they could have provided for public works programs for employment-creating opportunities for the poor in both rural and urban areas. And four, they could have further rationalized the tariff structure by levying regulatory duties on imports that feed consumption by the rich without doing much to increase investment in the economy. A quick glance at the budget gives the impression of a glass half full; it could have been filled a bit more to address the problems that continue to produce recurrent crises in the economy, but the policymakers chose not to follow that route.

Conclusion

Will the new government be able to address the many economic problems facing the country? This will need political resolve as well as careful planning. A small step in that direction was taken by the appointment of a panel of experts to assist the Planning Commission to come up with a programme of change and reform. The panel is made up of the best economists available, and they should be able to recommend a programme that focuses on structural reform. The fact that the Planning Commission has taken that step suggests that the new government is empowering the organization that was created for this purpose more than half a century ago. The Commission was overshadowed by the Ministry of Finance during the Musharraf period because the man who headed the ministry did not have the self-confidence to ask for advice. During his tenure, economic policymaking became ad hoc, subject to personal whims and pressures exerted by powerful groups of lobbyists.

What should the panel focus on in attempting to develop a programme? It must aim to achieve three goals. First, it must convince those interested in the economy that the country is serious about reform and development. Two, it must devise a plan to rescue the country from the economic meltdown it is currently experiencing. Three, it must put the economy on a trajectory of growth that is not only sustainable but would increase national income at a rate comparable to that of other large Asian economies. A high rate of economic growth is needed to provide employment to those seeking work, bringing women into the work force, and reducing interpersonal and interregional income disparities.

It always helps to focus on the positive when thought is being given to the development of a medium-term growth strategy. All the talk about current economic stress has diverted attention away from what are the positive features of the Pakistani economy. I would like to mention at least three of these. First is the agriculture sector, long neglected by the federal government's policies in favour of some other parts of the economy. I have held the view for a long time that Pakistan's policymakers should give a very high priority to agriculture. The sector should lead the rest of the economy, provide jobs in both rural and urban areas, and increase exports. The second advantage resides in the country's large population that should be educated and trained to become an asset rather than a burden for the economy. The third is Pakistan's location in the middle of the most rapidly growing parts of the global economy.

Endnotes

1. Shahid Javed Burki, "Time for Donors to Assist," *Dawn*, July 22, 2008.
2. Among the recent books on Pakistan's economic history, the most useful are Shahid Javed Burki, *Changing Perceptions, Altered Reality, Pakistan's Economy under Musharraf, 1999-2006*, Karachi, Pakistan: Oxford University Press, 2007; S. Akbar Zaidi, *Issues in Pakistan's Economy, 2nd edition*, Karachi, Pakistan: Oxford University Press; and Pervez Hasan, *Pakistan's Economy at Crossroads*, Karachi, Pakistan: Oxford University Press, 1998.
3. President Ayub Khan provided a detailed background of the relationship that was developed with the United States during his tenure in office in his autobiography. See Mohammad

Ayub Khan, *Friends, not Masters: A Political Autobiography*, London, UK: Oxford University Press, 1967.

4. The most detailed account of the way this relationship was built is in Steve Coll, *Ghost Wars: The Secret History of the CIA, Afghanistan, and Bin Laden, from the Soviet Invasion to September 10, 2001*, New York, The Penguin Press, 2004. Also see George Crile, Charlie Wilson's *War: The Extraordinary Story of the Largest Covert Operation in History*, New York, Atlantic Monthly Press, 2003.

5. Once again the most detailed account of this relationship is from the head of the Pakistani state during this period. See Pervez Musharraf, *In the Line of Fire: A Biography*, New York, Free Press, 2006.

6. For a recent story of the way the discipline of economics has developed, see David Warsh, *Knowledge and the Wealth of Nations: A Story of Economic Discovery*, New York, W. W. Norton, 2006.

7. This part of the report draws upon the first annual report published by the Institute of Public Policy, Lahore, Pakistan, headed by the author of this chapter. The report was written by a group of six economists that included Sartaj Aziz, Shahid Javed Burki, Aisha Ghaus-Pasha, Pervez Hasan, Akmal Hussain, and Hafiz Ahmed Pasha. See *State of the Economy: Challenges and Opportunities*, Lahore, Pakistan: Institute of Public Policy, 2008.

8. There was intense debate in the country about the incidence of poverty. While the government claimed that the incidence declined by 10 percentage points during the last 4 years of the Musharraf regime from about 35 percent to 25 percent (see Government of Pakistan, *Pakistan Economic Survey, 2006-07*, Islamabad, Pakistan: Finance Division, 2007). Some independent analysts thought that the decline was considerably more modest. See Akmal Hussain, "A policy for Pro-Poor Growth," presented at the seminar Pro-Poor Growth Strategies sponsored by the UNDP and Pakistan Institute of Development Economics, Islamabad, Pakistan, March 17, 2003.

9. *State of the Economy.*

10. Hafiz Pasha and Aisha Ghaus-Pasha, *Growth of Public Debt and Debt Servicing in Pakistan*, Research Report, No. 17, Karachi, Pakistan: Social Policy and Research Center, 1997.

11. This is not a new development. In fact, it was the subject of a book length study by Ishrat Hussain while he was at the World Bank. Hussain was an important player in the team assembled by President Pervez Musharraf to manage the economy. He was Governor of the State Bank of Pakistan, the country's central bank, from 1999 to 2005. See Ishrat Hussain, *Pakistan: The Economy of an Elitist State*, Karachi, Pakistan: Oxford University Press, 1999.
12. *State of the Economy.*
13. This theme was developed by the author in a series of articles written for *Dawn*, Pakistan's largest selling English language newspaper. See Shahid Javed Burki, "Agriculture: A Shift in Paradigm," *Dawn*, June 9, 2008; and "Agriculture: Public Policy Options," *Dawn*, June 23, 2008.
14. For a discussion of the problems in agricultural research, see *Punjab Economic Report, 2007*, Lahore, Pakistan: Government of Punjab, 2007.
15. For a discussion of the Indus Water Treaty, see Aloys A. Michel, *The Indus River*, New Haven, CT: Yale University Press, 1967.
16. The award was given by the NFC that I headed as Minister of Finance in the caretaker government that was in office from November 1996 to February 1997.
17. The new government constituted a new Finance Commission in May 2008.

Chapter - 5

Surviving Economic Meltdown and Promoting Sustainable Economic Development in Pakistan

—S. Akbar Zaidi

This chapter was originally meant as a comment on Chapter 4 of this book by Shahid Javed Burki, "Pakistan's Economy: Its Performance, Present Situation, and Prospects." However, given the very detailed and comprehensive nature of Mr Burki's arguments, I find myself with few disagreements, if any at all. Mr Burki has laid down a substantive manifesto and policy document with which few economists would disagree. Consequestly, my chapter, refers to Mr. Burki's as needed, but does not repeat any of his arguments, or the large amount of his data and facts with which I agree. My chapter builds on Mr Burki's thesis by raising issues that he did not discuss, although he has written about many of them quite extensively in the past. Both chapters should be seen as complementing each other rather than as responses or critiques and should be read in tandem. Where Mr. Burki's chapter provides ample economic data and analysis, my chapter enhances the debate by building on its economic base through the use of a political-economy perspective. While many of Mr. Burki's economic interventions are logical, appropriate, and timely, many will be confronted by vested and institutional interests that hamper any attempt to implement these urgently needed reforms that he proposes. Moreover, I argue that unless some key political issues, no matter how symbolic they may appear to be, are resolved, much of the substantive part of the economic policy will remain unimplemented, and the economic meltdown will become far worse, perhaps even threatening the democratic transition currently underway in Pakistan.

The Structural Transformations of Pakistan's Economy and Social Structure

Mr. Burki presents a number of key structural factors that have affected Pakistan's economy over the last 60 years and demonstrates how these shifts are affecting economic and social structures. In this section, I add a number of other factors that have changed over time and are going to affect how Pakistan's economy, its social sector, and politics respond to such dynamic influences.

Perhaps the most important factor that has undergone substantial change and transformation, as Mr. Burki has also pointed out in his chapter, one that, sadly, many other Pakistani social scientists still do not comprehend, is that Pakistan is neither a so-called feudal, agricultural, rural, or even traditional society or economy. Only those social scientists who write their chapters on anecdotal evidence still talk of Pakistan as being feudal. Even a cursory examination of any kind of economic data suggests that this is not so. With the agricultural share of the gross domestic product (GDP) falling drastically from 26 percent in 2000 to 20 percent in 2007, agriculture has lost its predominance in the economy. The share of agricultural labour has also fallen from more than half of the total in 1990 to 43 percent today, and land tenure relationships and landholdings have changed markedly. In terms of social values and behaviour, many analysts still refer to them as feudal, perhaps authoritarian, discriminatory, or undemocratic. However, these adjectives describe the nature of social relations between people, values, and behaviour in many highly developed countries as well. Therefore, if we are to understand social change and transformation, it is critical that we look beyond stereotypes, which will only limit our ability to observe and understand.

This is particularly so with regard to stereotypes such as Pakistan is an agrarian economy, and the view that Pakistan is largely rural. Raza Ali's extraordinary research on the 1998 Census[1] clearly showed that Pakistan was an urban country with perhaps 50-55 percent of the population living in settlements that by no stretch of the imagination could be called rural. A decade later, the forthcoming census will most certainly help us define the present social and economic relations of exchange and production and will reveal an even larger urban population. Moreover, with the increasing prevalence of communications technology of all sorts such as phones, electricity, roads, and other social services that are easily accessible, if not available, to so-called rural dwellers, the arbitrary binaries between urban and rural begin to fade. While much data can be shown to emphasize this point, the simple fact is that of the one million mobile phones added to the 81 million in service in Pakistan *every month*, the large majority are rural users, or those outside the geography that is administratively defined as urban. Pakistan is increasingly, if not now predominantly, urban, which engenders far more possibilities and opportunities to build on its economic and social structures. However, one must also recognize that there is the possibility that unless this urban transition is adequately managed, it could implode.

A particularly interesting consequence of this demographic change is how it affects Islamic political parties. Results from the most recent election—probably the fairest in Pakistan's history—show that for the most part, electoral Islam has been reduced to being primarily a rural phenomenon, Islamic political parties won largely in rural areas—from Balochistan and the Northwest frontier Province (NWFP)—and from the least developed districts in these provinces.

These structural shifts in economic and consumption patterns have finally given rise to the emergence, substantial growth, and consolidation of a Pakistani middle class. The consumer boom that has taken place in Pakistan over the last decade or so would not have been possible without the existence of a sizeable middle class. The exact size of the middle class is difficult to estimate or measure, and one hopes that some approximation of its size will emerge through research. On account of easy availability of credit, an argument can be made that supports the claim that a consumer-based middle class is the true stimulus of the economy. For instance, the numbers of cars and of motorcycles doubled in Pakistan between the period 2001-07; mobile phones, which had a density of just 5 percent of the population in 2004. By 2008, that percentage skyrocketed to 51 percent of the Pakistani population. Moreover, despite growing regional and income disparities, per capita income has almost doubled since 2000.

While an economic middle class clearly exists, one can perhaps surmise that along with the substantial growth in the news and information media, this class has also become more aware of its rights and perhaps even responsibilities. Perhaps it was these new emergent and assertive groups who participated in, and gave direction to, the political and civil society movements of 2007, of which Pakistan's media revolution played a key role. When the last elections were held in 2002, there was only one private TV channel in the country; today there are perhaps more than 30 private news and information channels broadcasting in all major languages. With constant information, analysis, and chatter about even miniscule political issues and developments, much of Pakistan's society has become involved and informed about what is happening throughout the country. While there are numerous rumours and justifications concerning a host of political stories, no one can any longer claim to be uninformed.

However, one must add a word of caution here. If the economic transformations from the agrarian, rural, and feudal structures have given rise to these new groups or the middle class, it is important to state that the political role of such classes need not be progressive, as is often incorrectly assumed and romanticized. The category of the middle class has no particular moral or ideological mooring. This group or class, can be as democratic and revolutionary as it can be fascistic.

Another factor that is affecting society and its relations is the increasing visibility of women in public, and not merely in Parliament. While the largest number of women have been elected from the General seats in the last elections, evidence from most urban population centers suggests that women are more visible at higher tiers of education, in the media, and in the growing service sector. It is not just that young women predominate in liberal arts and humanity colleges, but rough estimates suggest, for example, that in the case of Karachi University and the Government College University, women dominate the campuses by a huge margin, perhaps four-to192 one. While many observers point out that while on university and college campuses more women are certainly more visible, they immediately add that most wear some version of the *hijab,* suggesting a form of growing conservatism. These visual descriptions perhaps confirm the view of some that Pakistani society has become far more socially conservative, yet obscure the liberating element experienced by many of these women who have escaped from their oppressive, traditional, patriarchal, and familial bonds, if only for a few hours in the day. Clearly, just the fact that women are being educated in growing numbers and that they are working is a revolutionary transformation, which has multiple and diverse social, demographic, and economic repercussions that many would consider highly progressive.

A dramatic shift that has occurred in the last 6 years, and this might be the only benefit from the consequences of September 11, 2001 (9/11), is the substantial change that has taken place in India-Pakistan relations. On the one hand, little seems to have changed, with an inhospitable visa regime still in place, with bureaucrats trained in older schools of thought still determining relations between the two countries. While on the other, an astonishing set of figures paint a completely different picture. For example, *India is today Pakistan's seventh largest trading partner as a source of imports*, and the first three in this list primarily supply oil to Pakistan. Pakistan imports more from India than it does from France, Germany, Canada, Switzerland, Iran, Turkey, or even Thailand! Overall, *India is Pakistan's ninth largest trading partner*. In 2000, the official trade between the two countries amounted to approximately $235 million. Today, that number has grown to over $1.4 billion. And this despite challenging travel and visa restrictions for both traders and businessmen. One of the few positive policy actions taken by the new democratic government was the extended set of measures outlined in the Trade Policy for 2008-09 announced on July 18, 2008, in which trade with India was encouraged and a number of new concessions given. Early estimates suggest that if these opportunities are taken, trade with India will cross the $3 billion mark, an astonishing turnaround from just 8 years ago. In this case, India may become Pakistan's third major non-oil trading partner. The consequences of such a positive development is that it could substantially change Pakistan's political economy.[2]

In terms of investment interests, a new phenomenon is the emergence of business interests from the United Arab Emirates (UAE) and other Gulf States. Awash with excessive amounts of revenue from the rise in oil prices, Arab sheikhs have been buying up key sectors in Pakistan. They have invested in real estate, banking, telecoms, information technology, and in other service sector tie-ups. While it is still too early to judge, there are indications that the UAE is getting involved in Pakistan's economy and politics to the extent that it can influence decisions. Both Nawaz Sharif and Asif Zardari have had very close ties—business and personal—with many of the rulers in the Emirates, and both have lived in Dubai for long periods of time. Moreover, the November 3, 2007, closure (following General Musharraf's Martial Law) of the private channel Geo News which was based in Dubai, UAE, suggests that numerous arms of the Pakistani state also have close connections with the Emirates' sheikhs. If UAE business interests grow, and given the overlapping business, personal, and political relationships, one can be sure that financial capital from the Gulf will influence, or keenly follow, developments in Pakistan.

These are just a few of the many changes that are transforming Pakistani society, its economy, its politics, and its social relations of exchange and production. There are many reasons for these changes, from excess capital liquidity to globalization, to the media boom, to women's education, and similar trends. Some of these, such as trade with India, are reversible, but many suggest a more permanent state. There is a need for scholars to interpret and further explore such trends to examine and understand what, if anything, they mean for political transitions and transformations, and for economic development. One must understand that while there have been substantial and noticeable transformations, some institutions and some forms of politics have still not changed.

The Politics of Economics: The 2008-09 Budget

Mr. Burki gives a comprehensive assessment and critique of the 2008-09 federal budget announced on June 11, 2008, the first from the new government of Prime Minister Yousuf Raza Gilani. Mr. Burki ends his chapter by saying that while the new government had an opportunity and perhaps the responsibility to undertake a number of reforms, some of them urgently needed, the opportunity to do so was lost. One cannot agree more with Mr. Burki's assessment, and many economists have criticized the new government for not dealing with many of the critical issues that Mr. Burki has highlighted, ranging from exorbitant food prices, the rising price of oil, the cost of doing business, stabilizing the rupee, etc. In this section, I examine how the politics of the new government affected its ability to act, and I argue that it first had to address some important political issues before economic management could become the primary focus.

If the economy is in as bad a shape since March 2009 as the two Finance Ministers have maintained, then the budget presented in Parliament in June 2008 failed to address issues to turn the economy around. In fact, by not addressing pressing problems at this juncture, the government has made things far worse for itself within a very short space of time. This lost opportunity to move forward on the economy is symptomatic of the way this government has functioned, a fact displayed in the two other critical arenas with regard to the president and the judiciary, where it was unable to act and to move forward with deliberate speed. Its failure to act clearly, forcefully, and timely on the economic front will, in all probability, only add to causes that will result in its own undoing.

Of course, the present government is in no way responsible for the numerous problems that afflict Pakistan's economy, yet the fact that it is in power assumes that it must take responsibility if it fails to address these problems. While we celebrate the return of the democratization process, albeit one that continues to be partial and interrupted, we also expect the government to fulfil its responsibilities.

As Mr. Burki's chapter indicates, the main economic indicators show that a deteriorating trend has been in process for some months now. The GDP growth rate, expected to be 5 percent, is still considerably higher than the average for the 1988-2003 period, but is lower than the trend seen over the last 5 years. While perhaps this slowing of the growth rate was to be expected, given adverse international commodity prices and because previous growth was built on a weak foundation, expectations suggest that growth is going to be lower for some years to come. With the growing fiscal deficit which is approximately 7.8 percent of GDP, the highest in over a decade and that worsened as oil prices rose, all estimates for GDP growth seem to be overly optimistic. While these three key indicators need to be immediately addressed by the government, inflation is the single most important issue affecting all citizens, and this problem demands action from the current democratic and popular government.

With inflation at approximately 28 percent, the highest in the last 3 decades, any government would have had a major task dealing with causes that are not under its control. The global rise in food and oil prices are the main reasons why inflation is so high, although a number of poor decisions and an equal number of indecisions by the Shaukat Aziz government and the caretaker government have made things far worse. Moreover, the economic policies of the previous government were responsible for creating an artificial bubble, which resulted in a substantial deterioration in income distribution.

The Finance Ministry must realize the scale of the issues it must confront and quickly deal with them if it is to make a difference on the economic front and stop the economy from deteriorating further. However, the budget represents a substantial attempt to turn the economy around, and, although a couple of measures have been taken, they are insignificant compared to the nature and scale of the problems. The imposition of import duties on luxury goods was long overdue, and the attempt to provide an income support programme for the poor is welcomed. Development expenditure has also been raised, and one can only hope that better and effective targeted provision of all government expenditures will take place. However, a number of measures that were announced in the budget are troubling, and many are conspicuous by their absence.

The cut in subsidies might help to marginally lower the fiscal deficit, but will probably result in higher prices for oil, power, fertilizers, and food items for consumers, especially the poor. Similarly, an increase in the proportion of indirect taxes will also have a disproportionately higher impact on lower income earners. Neither of these two measures will help the poor and will further challenge their ability to survive under worsening economic conditions mostly related to rising inflation.

With a tax-to-GDP ratio of a mere 10 percent and with a fiscal deficit of 7.8 percent of GDP, one would expect the government to be considerably more imaginative on the revenue generation front. It is unforgivable that the government has allowed the exorbitant profits from the stock market to go untaxed for another 2 years, a concern also raised by Mr. Burki. There is no reason why profits from speculation should be untaxed under stable and normal circumstances, and under conditions where the economy is facing serious crises, such generosity is criminal.

The government should have gone out of its way to give inflation and food shortages its highest priority. It should have taken a short-term, immediate focus, which would have meant compromising on other issues at the moment, and a medium and longer-term economic agenda. One would have expected that the Peoples Party election manifesto, launched with much fanfare, would have had more substantive issues addressed with regard to the economy, some of which would have found expression in its first budget.

The democratically elected government in power today is struggling with an economic meltdown and economic crisis affecting its own constituency, but also a crisis of legitimacy, effectiveness, and political control. The disappointment one has with regard to the budget is just another indication of the democratically elected government's failure to act on a number of critical issues, both economic and political. Although a new president has been elected, it is quite clear that the new government in Pakistan seems vulnerable and weak. The war on Pakistan's borders seems to be getting out of hand, and differences with Pakistan's main ally, the United States, bode ill for any economic aid or rescue package that was envisioned. Economic and financial issues—such as the continuing deterioration of the rupee, the lack of foreign direct investment, the stock market unable to reflect any bullishness—are all linked to political stability in Pakistan.

Political Stability and Development.[3]

The arguments made in the previous section need to be reemphasized; without political stability in Pakistan, economic development will not take place. Political stability is a prerequisite for economic development. There are three arenas where political stability needs to be managed and resolved. First, there is the need for the newly elected democratic government to deal with the regime, institutions, individuals, and power structures in place since 1999, and to establish its writ. Second, there is a need to resolve provincial, domestic, and local political issues, such as the National Finance Commission (NFC) Award mentioned by Mr. Burki, inter- and intra-provincial disputes and inequalities, and issues about devolution and local government. In this category, there also is a need to develop a level of consensus between the different political parties vying for power in a more accommodative political framework. Finally, and equally importantly, there is a need to address and resolve issues related to Pakistan's neighbours, the war on terrorism, and in particular, the relationship between Pakistan's newly elected leaders and the U.S. administration. This last point is probably the most important and affects the other two, hence, it is to this I now turn.

It took the previous U.S. administration under President George W. Bush several months to come to accept the new reality in Pakistan after the 2008 elections and that its old and trusted ally since 2001 was not acceptable to the new government in Pakistan. For 6 months, it seems, the United States put pressure on Pakistan's new elected leaders to accept General (Retired) Pervez Musharraf as President of Pakistan, even after he and his political allies were resoundly defeated in the February 18, 2008, elections. The U.S. administration under President Bush, because of its own agenda and need in the region to fight and win the war on terrorism, had not been able to move beyond its political arrangements in Pakistan since 2001. By supporting General Musharraf after 9/11, almost unconditionally, the Bush administration was seen as the most powerful force propping up a military dictator in Pakistan. While political parties were equally to blame for not working for democracy and for being the General's collaborators and ensuring his longevity, it would be fair to say that General Musharraf continued to stay in power largely because of U.S. support for him. While the U.S. position was understandable, perhaps until the February 18 elections, the victory of democratic forces since then, and most importantly, the rejection of Musharraf's policies by the Pakistani electorate, not only delayed the democratic transition, but was also the main reason for political instability in Pakistan.

In the period between early February to the end of June, U.S. State Department officials had more meetings with the two most important unelected Pakistani civilians—Asif Zardari and Nawaz Sharif—than they probably had with Pakistan's quasi-civilian leadership between 2002-07. These meetings were held not just in Pakistan, but apparently in London, United Kingdom, and the Gulf States as well, whenever the two Pakistani leaders were abroad. Moreover, more substantive quantitative research would also support the impression that the number of meetings the two leaders had with the U.S. Ambassador to Pakistan and their excessive public presence in the media has seldom been as high as it was in the last 5 months before, during, and after the February 18 elections. Public announcements by senior State Department officials seemed to suggest that these meetings and the pressure put on the Pakistani leadership was to garner continued support for the U.S. war on terrorism in the region and to ensure that President Musharraf was not removed.

Rather than support the process of further democratization and Pakistan's emerging democracy to a more sustainable level, the United States continued to support the one man who was unpopular and probably illegally in power. The needs and requirements of the previous Bush administration and the desires of the Pakistani people were seen to be at odds. Eventually, however, President Musharraf was replaced by the leader of the

Pakistan Peoples Party (PPP), Asif Ali Zardari, who was sworn-in as Pakistan's democratically elected President. Only time will tell if President Zardari is able to gain the same trust and confidence of the new U.S. administration under President Barack Obama, as the man he replaced enjoyed with the Bush administration. With the United States playing such a crucial role in the region and particularly in Pakistan, and with democracy still emerging and establishing itself in the country, the relationship between Pakistan's elected leaders and the new U.S. administration under President Obama will be crucial. Moreover, it is also important for President Zardari and his new government to diversify and expand their diplomatic relationships with other regional powers, such as China and India, and to build alliances that look beyond too much reliance and dependence on the United States.

It is important to point out that U.S. intrusion in the region and on Pakistan's borders in military terms does not bode well for Pakistan's government and its democracy. Military attacks by U.S. drones on the border region with Afghanistan have already caused a great deal of resentment in Pakistan and help to increase anti-U.S. sentiment. Moreover, statements by Pakistani leaders, both civilian and military, seem to be at odds with those of U.S. military leaders, and clearly, differences of opinion, priority, and tactics have emerged. This relationship between the present U.S. administration and the new democratically elected leadership in Pakistan, and its relationship through the war on terror, may determine far more than just the future of democracy in Pakistan. Much more is at stake, and a misreading or mishandling of the situation will have multiple consequences, most of them ugly, on numerous actors and relationships in the region.

Conclusion

In agreement with Mr. Burki's suggestions of how Pakistan should exploit its many positive economic trends, my chapter has only added to the debate, suggesting that some key political and diplomatic issues need to be addressed quickly, alongside issues related to the economic meltdown affecting the economy and the Pakistani people. Political instability and factors related to the war on terror will hinder economic development. It is in the interest of the new U.S. administration under President Obama to support and strengthen the civilian democratic setup in Pakistan and to show a longer-term commitment to stability in the country. The longer any uncertainty lasts, the greater the economic meltdown. The blowback of an economic crisis, along with a political one, will benefit neither the people of Pakistan and nor the U.S. administration in its war on terrorism in the region. Continuing instability and the unravelling of the Pakistani economy and the state are likely to have consequences far beyond Pakistan itself.

Endnotes

1. Raza Ali, "Underestimating Urbanisation?" in S. Akbar Zaidi, ed., *Continuity and Change: Socio-Political and Institutional Dynamics in Pakistan*, Karachi, Pakistan: City Press, 2003.
2. For domestic linkages and political-economy consequences of trade with India, see Chapter 20 of my *Issues in Pakistan's Economy*, Second Ed., Karachi, Pakistan: Oxford University Press, 2005; my *Pakistan's Economic and Social Development: The Domestic, Regional and Global Context*, New Delhi, India: Rupa and Co., 2004; and *South Asian Free Trade Area: Opportunities and Challenges*, Washington, DC: U.S. Agency for International Development (USAID), 2006. I must also acknowledge that Mr Burki was one of the first Pakistani economists to talk and write publicly about improving Pakistan's economic relations with India.
3. See my unpublished background paper presented at the International Debate Education Association (IDEA)/Centre for Security and Defence Studies (CSDS) Conference entitled "Democracy, Development, Dictatorship and Globalization: The Complicated Histories of Pakistan," held in New Delhi, India, June 17-18, 2008.

Chapter - 6

Pakistan 2020:
The Policy Imperatives of Pakistani Demographics

—**Craig Cohen***

Pakistan poses a unique challenge to U.S. foreign policy. The government has been a front-line partner in the Bush administration's War on Terror, but is also home to the Taliban, al Qaeda, and remnants of the nuclear proliferation network of A. Q. Khan. The United States depends on Pakistan's cooperation, but its people and government remain wary and at times hostile toward the United States. Even Pakistanis sympathetic to U.S. goals often call for greater patience on the part of Washington, but Americans are unlikely to become disinterested observers in Pakistan any time soon.

One of the toughest short-term challenges facing the next U.S. administration is how to address the problem of militancy on Pakistan's western border. Success in Afghanistan and security at home depend on finding effective solutions in Pakistan's Tribal Areas. Success is likely to remain elusive as long as Pakistan and the United States remain on different time horizons. The United States feels the urgency of the threat, while Pakistanis take a longer-term view of progress. Only months after the February 2008 parliamentary elections, Washington became frustrated with the weakness of Pakistan's new civilian government and its unwillingness to address the problem of

---

* The author wishes to thank Tara Callahan, who provided research assistance for this chapter.

militancy head-on. For its part, many in Islamabad see containing the militants as a viable option, and incorporating the Federally Administered Tribal Areas (FATA) as a generational task only made more difficult by direct U.S. action.

This is a time of great unpredictability in Pakistan and for U.S. decisionmakers. Major questions surround Pakistan's leadership, economic future, and social stability. Could Pakistan become a steadfast ally of the United States? Or will the United States find itself in direct military confrontation with Pakistani forces? Is Pakistan sliding toward collapse? Does it pose a clear and present danger to the United States? If so, what are the policies that could mitigate this threat? With Pakistan, everything is on the table, from billions of dollars of U.S. aid to Predator missile strikes and pariah status. It is no wonder that U.S. policy toward Pakistan has been trapped in a short-term mindset since September 11, 2001 (9/11).

Forecasting Pakistan's near-term political future may be a fool's errand, but anticipating trends that will shape its evolution over the long-term is possible and necessary. A closer look at Pakistan's demographic challenges raises a number of policy imperatives for Pakistan's government and the United States. Both will have to contend with an ongoing demographic transition characterized by a shifting age structure and migration pattern that are likely to place newfound resource pressures on food, water, and energy, and heighten the importance of addressing poverty, education, and violence.

Population trends are not destiny. They simply present new challenges and opportunities for governments and outside actors. How well the Pakistani government and the United States recognize, adapt, and get ahead of these trends will shape the Pakistan that will emerge in the years to come.

Pakistan's Demographic Future

A chapter on demographics may seem out of place in a book on Pakistan's nuclear future. What do population trends have to do with nuclear weapons? One possible way to think of the correlation is that nuclear weapons are the deadly tip of the iceberg, while demographics are the danger lurking far below the surface. Demographic visions traditionally have alerted us to external threats that could have destructive consequences for our own society.[1] Demographic projections have become a sort of "geopolitical cartography" for national security planners, helping them to avoid dangers hidden in the future's hazy unknown.[2] Since the earliest days, demographics have been viewed as a determinant of other societies' hostile actions, capacity, and intent, and the nuclear age is no exception.[3]

The size, security, and possible use of Pakistan's nuclear arsenal in 2020 will be a function of individual decisions by Pakistani leaders and its national security community. These decisions, however, will be shaped by a broader domestic and international context. Demographics will play an important role in determining this context, helping to shape Pakistan's politics, social cohesion, and economic growth. The demographic effects will be indirect, and they will operate on a longer timeframe than any democratic political calendar. Demographic change, in the words of one recent study, "shapes political power like water shapes rock. Up close the force looks trivial. But viewed from a distance of decades or centuries it moves mountains."[4]

Population has always been linked to security. Traditionally, the size of a body politic has been leaders' main demographic concern: the larger the population, the greater a society's wealth and power. In the most elementary sense, a larger population provides more men to field in battle. Pakistan had just under 40 million citizens in the years immediately following partition, but today it has somewhere in the neighbourhood of 170 million people and is the seventh largest country in the world. Pakistan's population doubled between 1961 and 1982, a period of just 21 years.[5] The United Nations (UN) projects that by 2050 Pakistan's population could double again to more than 350 million people, making it the world's third or fourth most populous country.[6] One would expect a Pakistan of 350 million people to wield significant influence on the world stage, particularly in the Muslim world. The question of whether Pakistan can become a global or even a regional power, however, is very much tied to its stability and economic growth. India, its main strategic rival, already dwarfs Pakistan with over one billion people. It is expected by some to grow to more than 1.6 billion by 2050, overtaking China's own growth projections for this period.[7]

Population growth is not always a positive occurrence. As far back as Malthus' writing in the late 18th and early 19th centuries, commentators have worried about populations outstripping their environments. Theories of Social Darwinism have been discredited in most circles, but many leaders today recognize that high population densities and high rates of population increase can undermine gains from economic growth and potentially contribute to social and political instability. Three factors determine population: fertility, mortality, and migration. For its part, Pakistan has successfully curbed its fertility rates after decades of effort. Pakistan first implemented an anti-natalist policy in 1965, but it was not until the 1990s that it experienced a fertility downturn. Since the 1960s, Pakistan's population grew at a staggering rate of close to 3 percent per year. Fertility rates in the 1970s and 1980s hovered between six and seven births per woman. Fertility began to decline as families migrated to urban areas, women married later, and family planning became more accepted practice. During this decade, fertility rates stand around four births per woman in Pakistan.[8] By 2050, some expect the total fertility rate to fall to between 1 and 3 births per woman.[9]

Falling fertility rates has meant that Pakistan is presently undergoing a demographic transition. Demographic transition explains the shift from the high death rates and high birth rates of a preindustrial society to the low birth rates and low death rates of industrialized economies. Pakistan's crude death rate declined progressively from 24 deaths per 1,000 in 1950 to 8 deaths per 1,000 in 2006.[10] Pakistan's death rate declined during this time at a much faster rate than its fertility rate. The result has been a shifting population demographic away from the classic pyramid model to a more cylindrical shape. The main reason the structure of population aging matters is that there is a "mismatch between the timing of human productivity and human consumption."[11]

Because of the time lag between changes in fertility and changes in mortality, Pakistan is experiencing the possibility of what is called a "demographic dividend." The potential for a demographic dividend occurs when a lowered birth rate leads to changes in the age structure of a population. In this case, an increase in working age population and decline in dependent age population results in economic gains.[12] Simply put, young people require a society's investments in health and education, working-age adults supply labor that fuels this investment, and the elderly again require investments in health. A country with a large labor supply relative to its young and old has the potential to realize significant economic growth because the dependency burden is low. Demographic dividends create the possibility for economic growth by "improving labor supply, increasing savings, and allowing development of human capital."[13] Pakistan's median age in 2006 was 20 years. By 2050, it is projected to be 33 years.[14] The proportion of Pakistan's working age population of 15 to 64 reached 59 percent in 2006.[15] Capitalizing on the possibility of a demographic dividend is currently up to the Pakistani government.

This dividend can only be realized in the right policy environment such as occurred in South Korea over the second half of the 20th century or is occurring in China and India now. Studies have demonstrated that as much as one-third of East Asia's economic miracle can be attributed to a demographic dividend.[16] What Pakistan is currently experiencing is a once in a lifetime opportunity as the working age swells and dependency ratio declines. Demographers believe that Pakistan's window of opportunity probably opened in 1990 and is likely to close by 2045. The critical question is whether the labour market will be able to absorb an influx of new workers. As one commentator has asked, "Would these teeming numbers be actually a 'dividend' or would they be more of a threat?"[17] By 2030 Pakistan is estimated to have 175 million potential workers, 85 million of whom could be women. Realizing a demographic dividend is closely tied to female education and empowerment, as it depends upon sustaining lower fertility rates.[18] By 2050, the number of potential workers is expected to rise to 221 million.[19] If Pakistan fails to adequately train and educate its labor supply and grow its economy to provide jobs, difficult times could be ahead.

The other side of the demographic dividend coin is demographic danger. The most commonly discussed demographic threat is known as "youth bulge." This is the period typically before the demographic dividend can be realized when the huge tide of young people has not yet entered the labour market. This creates enormous pressures on the state to provide health, education, and other services. As this population becomes adolescents, the theory holds that single teenage men without the discipline of a good public, private, or military education and without the prospects of employment are more likely to engage in violence directed against the state and other groups in society, or engage in terrorism. Many have looked to the failures of the state education system in Pakistan, for instance, as a primary reason for the greater role *madrassahs* have played in educating young Pakistanis today. *Madrassahs* do not necessarily produce terrorists, but they do play a role in proselytizing an anti-modern, anti-Western world view.

There is no guarantee that Pakistan has weathered its period of youth bulge as it transitions to its dividend period. One recent study has argued that, similar to the aftershocks of an earthquake, "echo booms" reverberate every 2 decades after periods of booming fertility which are followed by a steep decline. This would mean that the number of Pakistanis between the ages of 15 and 24 would grow from roughly 7 percent of the population between 2005 and 2020 to over 30 percent between 2020 and 2035.[20] This would create a new period of stress for both state and society. Demographic transitions ultimately reduce threats of violence and instability, but these transitions proceed unevenly. It is in the midst of the transition—when inequality is growing, urbanization and migration are high, and contact with the global marketplace is on the upswing—that political instability and authoritarian reactions are most likely.[21]

Other than fertility and mortality, urbanization is the third demographic effect that shapes the contours of a country's population. Urbanization is a form of migration: citizens migrate internally from rural to urban areas, often in search of jobs and a better life. Migration has been an integral and often painful part of Pakistan's history. Four main migratory waves have shaped Pakistan's demographics: at partition from India; the war in neighbouring Afghanistan; workers' migration to the Gulf; and urbanization, including the growth of the megacity Karachi. The continuing effects of each are likely to play an important role in Pakistan's economic, social, and political stability for decades.

Between partition in August 1947 and the end of open borders in 1951, 6 million non-Muslims moved from Pakistan to India, and 8 million Muslims moved from India to Pakistan.[22] Most of the migrants to Pakistan were East Punjabis who settled in Punjab. Twenty percent, though, were so-called Muhajirs, Urdu speakers who settled in Sindh and had a significant influence on provincial and national politics.[23] During the 1980s, there was a comparable influx of people into Pakistan on account of the Afghan war. More than 2.5 million Afghans fled to Pakistan to escape the violence, settling primarily in Peshawar and Quetta in tight kinship and tribal networks.[24] Many of these Afghans still remain, despite large scale repatriation efforts after the fall of the Taliban in 2001.

Migration from Pakistan to the Gulf states took off in the 1970s as a construction boom drew workers from uneducated, rural areas of Pakistan. Skilled workers later followed. Savings sent back to Pakistan in the form of remittances have constituted the largest single source of foreign exchange earnings for Pakistan.[25] While some see remittances as a major financial resource that could be harnessed for development, the long-term effects of remittances on structural poverty are less clear.[26] According to the Pakistan Ministry of Finance, the total remittances sent to Pakistan between FY2002 and FY2006 were $4.57 billion. The United States was the single largest country source, although the Gulf states, if lumped together, provided the largest single regional total.

Pakistan has traditionally been a rural, agricultural country. In 1951, 83 percent of Pakistanis lived outside of cities and towns. Today, this number has fallen to 68 percent or less. Urbanization is progressing at a rapid 4.9 percent per year, and Pakistan is projected to be predominantly urban by the next decade.[27] More than half the urban population of Pakistan lives in the eight largest cities, and Sindh is the most urbanized province in Pakistan on account of Karachi. Rapid urbanization and the ensuing high congregations of people living in slums create a host of pressures on state and society. In Lahore, Pakistan, for instance, there are 6,500 sanitation workers for 7.5 million people. In Delhi, India, by comparison, there are 46,000 sanitation workers for 11 million people.[28] Urbanization also erodes traditional social structures and exposes migrants to "the social and cultural crosscurrents of modernity."[29]

The effects of these four migratory waves will continue to shape Pakistan. The pace of urbanization will create new strains and opportunities that could serve as an engine of industrialization and modernization, or else be captured by unstable and violent crosscurrents. The large presence of Afghans in western Pakistan continues to blur the Durand line separating the two countries and further complicates efforts to tie the Northwest Frontier Province (NWFP) and Balochistan more squarely to Pakistan's center. Muhajirs continue to play an important role in Pakistani and Karachi politics. Muhajir-Sindhi ethnic violence like what occurred in the 1990s remains a continuing possibility in Karachi. The Gulf has proved an important source of capital for Pakistan, but it is unclear what sort of political influence the transfer of wealth will have on Pakistan over time.

The critical point to note is that while Pakistan has undergone remarkable changes over the past 60 years, perhaps none have been greater than what has occurred over the past decade. Agriculture is no longer the lone driver of the economy. Automobiles, mobile phones, and an independent media have connected Pakistanis to each other and to the rest of the world in revolutionary ways. Women play a greater role in public life, including at universities. Pakistan is continually renegotiating its relationships with Islam, India, China, the United States, and the Gulf.

Understanding these changes and the effects of Pakistan's demographic transition are vital to understanding Pakistan's future. The country's ability to successfully weather this transition period will depend on two primary factors. The first is how its leaders manage to address the instability caused by increased resource pressures on food, water, and energy. The second is how the country manages to address three societal ills likely to be heightened by the transition period: poverty, lack of education, and violence. This chapter will discuss each of these dynamics before making a case for why U.S. decisionmakers ought to pay attention to Pakistan's long-term future and what policy options exist to mitigate peril. The United States must help Pakistan through this demographic storm, or else risk its worst effects washing up on our shores.

Food, Water and Energy

Population change is closely tied to resource availability. Food, water, and energy are all basic requirements for life and economic activity. Pakistan, like many countries, faces severe constraints on all three, and the potential for shortages is only likely to grow as populations increase. Pakistan was hit particularly hard by the global food crisis this year. The World Food Program reported that as many as 60 million Pakistanis were "food insecure" as a result of the global rise in commodity prices. Despite 6 years of sustained economic growth, roughly a quarter of Pakistan's population still lacks potable water.[30] The United Nations Development Program (UNDP) Human Development Report has predicted a global water crisis by 2025 that Pakistan is unlikely to escape.[31] Energy shortages continue to be endemic in Pakistan. This past summer, Pakistan's government set the nation's clocks forward by 1 hour to ease energy demand. These pocketbook issues traditionally have led to political instability in Pakistan, but few political leaders have been able to devise a long-term countrywide strategy for addressing food, water, and energy insecurity.

The global food crisis hit the world this year with "alarming speed, force, and depth," presenting a humanitarian, development, and strategic threat to countries around the world.[32] The price of basic foodstuffs skyrocketed as a result of high energy prices, increased demand from rising middle classes in China and India, the increased production of biofuels, poor weather potentially linked to climate change, and more systemic problems in agricultural production, trade, and the delivery of food relief.[33] Since the beginning of 2006, the average world price for rice has risen over 200 percent, milk by 170 percent, wheat by 136 percent, and maize by 125 percent. The new Pakistani government was forced to place a 15 percent export duty on wheat and to import millions of tons of the country's main staple in order to address the shortages. The UN World Food Program (WFP) estimated that close to 40 percent of Pakistan could no longer afford the poverty-line intake for food. Urban areas were hit particularly hard as food prices increased.

What is unfortunate is that Pakistan was near food self-sufficiency for wheat in the early 1980s.[34] Pakistan's emerging food security problem has been linked closely to the unprecedented increase in population.[35] As early as the 1990s, food projections were showing that the demand for rice and wheat would soon outstrip supply.[36] This is despite a relatively successful record of agricultural growth. Pakistan has always at its heart been a rural agricultural society, even though it is becoming increasingly urban. Agriculture still accounts for one-quarter of Pakistan's gross domestic product (GDP) and employs almost one-half of its labour force.[37] Seventy percent of export revenue stems from agriculture, and over one-half of industrial production comes from agricultural business.[38] A lasting agricultural crisis in Pakistan will have far greater implications than its effect on individual families. It is likely to severely impact the country's economic growth.

One of the main challenges to increasing agricultural production in Pakistan is low productivity and reliability of water. The total irrigated area of Pakistan increased by 80 percent between 1960 and 2005, from 10.4 to 18.8 million hectares. Upwards of 80 percent of Pakistan's cropped area is currently irrigated.[39] Farming is a water-intensive pursuit, taking 1,000 tons of water to grow one ton of wheat and 2,000 tons of water to grow one ton of rice.[40] According to one study, because of water shortages, Pakistan will be forced by 2025 to import large quantities of wheat amid "faminelike conditions."[41] That day may arrive sooner than expected.

At the root of the problem is that human populations continue to grow, but the amount of fresh water stays roughly the same over time.[42] Pakistan had essentially the same annual renewable water availability for its 35 million people in 1947 as for its 170 million people today. In 1981 there was close to 3,000 cubic meters of water for each Pakistani each year. By 2003, the number had fallen below 1,500, and by 2035, it is projected to fall below 1,000, the baseline that indicates water scarcity.[43] The Indus River basin covers 70 percent of Pakistan's territory. Its flow depends on melting snow, and its irrigation potential is thus limited to the months between May and September. The rest of the year, Pakistan depends upon stored water. The storage capacity of water reservoirs in Tarbela and Mangla are decreasing due to erosion from farming techniques that increase sediment and displace water.[44] If climate change reduces the snowcap on the Himalayan and Hindu Kush mountain ranges that feed the Indus, the river's flow could shrink even further. The lack of clean water impacts sanitation. UNDP estimates that almost 40 percent of Pakistanis lack adequate sanitation, increasing the spread of waterborne disease.

The Government of Pakistan is aware of what is required to address the country's water shortfall: new dams that can create new water storage facilities, more efficient farming techniques, and updated storage and irrigation systems. The government has unfurled a string of strategies, action plans and projects, but many observers are left with the sense that there is not a single, comprehensive plan to tackle the water crisis in Pakistan. At least one development bank has argued that without the creation of three new dams in Pakistan by 2016, the country will experience a severe water shortage by 2020. The politics of dam building in Pakistan are extremely complicated though, with each provincial government fearing that it will somehow be shortchanged. It is worth noting that General Pervez Musharraf's efforts as the country's de facto military ruler to move forward with dam construction in Kalabagh, Punjab Province, Pakistan, met with such stiff political resistance that he was forced to back down in 2006. It took 30 years for the four Pakistani provinces to agree on the 1991 water apportionment accord following the 1960 Indus Waters Treaty, and they are still fighting today.[45]

Water in Pakistan is vital to energy production. Electricity production from fossil fuels and nuclear energy requires huge quantities of water for cooling and other purposes, and one-half of Pakistan's electrical energy is hydro-generated.[46] Water is also essential for all other types of energy production. Future investment in alternative energy sources such as biofuels will further depend on irrigated land. At present, Pakistan receives roughly half of its energy supply from natural gas, 30 percent from oil, 11 percent from hydroelectric energy, 8 percent from coal, and 3 percent from nuclear energy.[47] Only 20 percent of oil demand is met from indigenous sources. In fact, Pakistan's dependence on imported energy is expected to increase considerably in the near to medium term.[48] This could be particularly damaging to Pakistan's economic situation given the soaring prices of energy.

Pakistan is currently experiencing an acute energy shortfall. Some have estimated that Pakistan is meeting only one-fourth to one-third of its power generation needs. Forty percent of households in Pakistan lack electricity and 18 percent of households have no access to pipeline gas.[49] There is a close linkage between power generation and economic growth. As the country's economy grows, its power generation needs will increase at a proportionately higher rate. Pakistan's energy deficit could double by 2025 if growth continues at its present pace.[50] Energy expansion, in turn, could lead to higher economic growth, and energy shortages could retard the growth process.[51] To meet rising demand, the government in the short term has sought to tap unexploited coal reserves in the Thar Desert in the Sindh Province of Pakistan and hydroelectric power from the north, both of which present serious logistical constraints. Over the long-term, Pakistan hopes to become an energy corridor between the Middle East and Central Asia.

The majority of Pakistan's natural gas production comes from Balochistan, a province that has engaged in a longstanding, low-level insurgency against Islamabad for decades. Pakistan's energy insecurity and its search for assured access to hydrocarbon resources has "magnified the economic and strategic importance of the province,"[52] as Balochistan accounts for almost 40 percent of Pakistan's natural gas production. The province—sparsely populated and underdeveloped—consumes under 20 percent of this production, though, and receives a "deficient share of revenues from the government's sale of natural gas," a main grievance of the insurgents.[53] The province's potential as a transit point for gas pipelines running between Iran and India and from the new Gwadar port to Central Asia increase its strategic importance. Balochistan stands as a good example of the way energy has fundamentally affected political stability in Pakistan over recent decades.

Pakistan's leaders must develop effective policies for addressing the instability likely to be caused by increased resource pressures on food, water, and energy during the country's demographic transition. Addressing the consequences of rapid urbanization is critical. When severe drought led to a 40 percent decline in wheat production in Sindh in the late 1990s, rioters stormed Karachi to protest the food and water shortages.[54] Rural poor often lack the ability to politically or violently mobilize in the way that urban populations do. As Pakistan's pace of urbanization continues, we are likely to see more rather than fewer disturbances in its major urban areas. Street protesters provide a unique challenge for the Pakistani military, which is relied upon to keep order, but which realizes that blood on the streets tarnishes its image as guardian of the state. It is possible that sustained shortages in food, water, and energy could lead to a decreased capacity of the Pakistani state to govern, increased migration, and civil unrest particularly in urban settings. The degree of instability will largely be a function of how the country manages to address three societal ills that could be heightened by the transition period: poverty, lack of education, and violence.

Poverty, Education and Violence

Pakistan in 2020 could very well become a wealthier, better educated, more stable society. If it can reduce its sources of violence and instability, attract foreign investment, provide government services, produce new jobs, and develop its human capital, Pakistan will have taken advantage of its demographic dividend, and allay U.S. concerns about Pakistan's nuclear future. Demographic trends may be robust predictors of population growth and movement, however, future availability of resources is largely a known entity, but future levels of poverty, education, and violence depend almost entirely on human decisionmaking. It is impossible to accurately forecast whether Pakistan's government will make the right choices and how external events may shape and impact those choices. What is possible is to provide a baseline assessment of the current state of these critical drivers of conflict, instability, and extremism, and then analyze how demographic pressures may impact these drivers in the years to come.

Poverty, in and of itself, is not a cause of conflict, instability, or extremism. It is, however, a phenomenon that shapes how communities and individuals perceive their future and the opportunities that will exist for them and their children. Relative poverty is more likely to produce the alienation, isolation, and grievance that prove fertile ground for political instability, internal conflict, and extremist sentiment. Countries undergoing a demographic transition are more likely to experience higher levels of income inequality as societies move from developing to industrialized economies. There are economic winners and losers in any society, but the winners and losers tend to move further apart during these transition periods. Government programmes and foreign aid may help to ease the burdens of those who suffer most during this transition, but it is ultimately economic growth that has the potential to lift millions of people out of poverty, as the world is witnessing in China, and to a lesser extent, India.

Pakistan's recent economic growth is well-known. From 2002 to 2008, the country's GDP grew by an average of 7 percent per year and per capita income increased by 5 percent per year, the highest rate in Pakistan's history.[55] Total investment reached 23 percent of GDP in FY2007, and foreign direct investment reached $5.1 billion, or 3.7 percent of GDP in FY2006.[56] Pakistan has negotiated trade agreements with China and a number of its neighbours, even if regional trade with India has remained mostly stagnant. Few could argue that the economic turnaround of this decade has not been beneficial for Pakistan. Pakistan's economic leadership during the 1990s was plagued by corruption, public debt, high deficits, and high poverty.[57] Musharraf's government brought macroeconomic stability and helped to deregulate key industries. The question, however, is whether in recent years Pakistan has experienced what William Easterly has called Pakistan's experience of the 1960s and 1980s: "growth without development."[58]

In 2006, six million families in Pakistan were still below the poverty line.[59] Pakistan's score in the UNDP's Human Development Index is currently out of 177 countries, sandwiched between Ghana and Mauritania.[60] There is widespread sentiment in Pakistan that the benefits of this decade's economic growth failed to trickle down to the majority of the population. Recently, there has been an additional worry that despite GDP growth, the country stands at the precipice of a major financial crisis stemming from the economic policies of recent years. Shahid Javed Burki, for instance, has argued that the recent GDP growth may have been artificial and is unlikely to be sustained without higher rates of investment in key industries like power generation and job creation for the rural poor.[61] A balance of payment crisis looms in Pakistan, and many fear a return to International Monetary Fund (IMF) lending.

Even if growth can be sustained, though, investments in health and education remain well below what is needed to realize Pakistan's demographic dividend. The success of this dividend will depend in large part on the country's ability to provide young people with the skills they need to succeed in the global marketplace. This means producing a work force that is globally competitive and can help Pakistan diversify from traditional agricultural-based industry like textiles. Without serious education reform, Pakistan is looking at large numbers of unemployable adolescents with few economic prospects who are sure to be the prime targets of those seeking to mobilize them for violent purposes.[62]

The UNDP Human Development Report in 2005 gave Pakistan the lowest score for its education index of any country outside of Africa.[63] Pakistan's overall literacy rate hovers between 40 and 50 percent.[64] For women, the literacy rate is below 30 percent, and for women in the Federally Administered Tribal Areas (FATA), it is only 3 percent. Pakistan's primary school enrollment rate in early 2000 was the lowest in South Asia.[65] In 2005, Pakistan's secondary school enrollment stood at just 27 percent of eligible students and less than 5 percent went on for tertiary education. Male children in Pakistan receive an average of 3.8 years of education, while female children receive an average of 1.3 years.[66] A host of problems plague education in Pakistan: internal mismanagement, poor quality textbooks, ghost schools, shoddy infrastructure, and discrimination. The single greatest challenge to reforming education in Pakistan is the poor quality of its teachers, who lack skills and incentives and who often fail to show up for work because of their low salaries. The result is that more Pakistanis are turning away from public education to attend private schools and *madrassahs*.

Much of America's attention on education in Pakistan has focused on the role of *madrassahs*. The linkage between *madrassahs* and terrorism is tenuous, however. While militant recruitment does take place in *madrassahs*,[67] it is probably more likely, as many have suggested, that an al Qaeda commander has graduated from the London School of Economics than a Pakistani *madrassah*. Still, these schools fail to educate Pakistanis in a way that will make them competitive in the global marketplace. They also contribute to an environment in which anti-modern and anti-Western views are more likely to take root. For those looking for a more moderate and tolerant Pakistan to emerge, the answer is unlikely to reside in *madrassahs*, though there is no guarantee it will reside in Pakistan's public education system either. Public schools in Pakistan continue to provide textbooks with historical inaccuracies based on religious animosities rather than historical, scientific, or economic explanations.[68] The problem for the United

States, though, is that efforts to try to help the Pakistani government address curriculum and textbook challenges touch a third rail of sovereignty in Pakistan and is sure to provoke significant backlash.

The U.S. focus on *madrassahs* is in effect a search for a simple explanation of the roots of violence and extremism in Pakistan. Unfortunately, there is no single structural factor that one can identify. The general absence of rule of law in Pakistan means that the police are viewed by most citizens as predators rather than protectors. Strong secessionist feelings and sectarian and ethnic tensions tend to overwhelm weak political institutions that have been purposefully kept weak by military rule. Regional and great power pressures from India and the United States tend to negatively influence stability. The country is awash with small arms. In such an environment, Graham Fuller's great question takes on a profound importance: "who will be able to politically mobilize this youth cohort most successfully: the state, or other political forces, primarily Islamist?"[69] The potential exists in complex tribal environments like FATA for the emergence of an outside entity with "powers of oratory and organization" who, with the assistance of outside money, can lead a revolt against traditional authority.[70]

Young people have been playing an increasingly important role in militant organizations in Pakistan today.[71] These militants have killed hundreds of tribal chiefs and upended traditional authority in FATA, making it less likely a tribal uprising will succeed in casting out groups like al Qaeda. The interwoven web of militant organizations in Pakistan works to al Qaeda's benefit. Al Qaeda has no dedicated recruiting infrastructure in Pakistan, but relies upon this informal network.[72] Historically, militant recruitment has revolved around the Indo-Pakistani conflict and has taken place out in the open, but since 9/11 it has gone underground and has tended to use anti-U.S. sentiment to motivate new cadres.[73]

If Pakistan is unable to sustain its economic growth, in part because of rising resource pressures, the country in 2020 could have millions or potentially tens of millions of unemployed young people who have not been properly educated to compete in the globalized economy. This will be a population that came of age during the War on Terror at a time of great antipathy toward the United States. Even if rural areas in Punjab and Sindh remain relatively quietist traditional societies as they have for decades, the increasingly populated cities and the heavily trafficked border regions will have access to networks of influence around the world. The Gulf, with the rising importance of its Sovereign Wealth Funds and growing source of remittances returning to Pakistan, is likely to have a heightened political influence. Today's interconnected world means that vectors of prosperity can quickly become vectors of instability.

Understanding the Risks

It is worth asking why Pakistan's future—its demographics, growing natural resource pressures, and efforts to address social ills—should matter to the United States. After all, the short-term dangers in Pakistan are numerous and challenging enough to suck the oxygen out of any long-term policy discussion. Furthermore, many countries around the world struggle with similar long-term challenges and sustain normal partner relationships with the United States. Why can't the U.S. Government continue to focus on short-term challenges in Pakistan while supporting the traditional programmes to promote good governance and economic growth?

The answer is that Pakistan may be the country where nuclear risk is greatest—where nuclear material is least secure, terrorists most active, and nuclear exchange with a neighbouring state most likely. The longterm stability of Pakistan's state and society matters to the United States because the consequences of nuclear terrorism or a nuclear war could be catastrophic to the region and to American interests and lives.

The nuclear experts who study Pakistan tend to downplay the nuclear threat, but the potential for nuclear war, nuclear theft, or nuclear accident will increase in Pakistan as domestic instability increases. Pakistan's safeguards against these nuclear risks—the military's cohesion and professionalism, established command and control procedures, a robust conventional response capability that reduces the potential for nuclear use, a politically moderate and generally pro-Western government and military leadership—all could erode or disappear in the years ahead as demographic pressures rise and the fabric of Pakistan's state and society come under additional strain.

Pakistan's relations with India, for instance, have thawed since the 1999 Kargil War, but India continues to dominate the thinking of Pakistani national security planners. India's growing presence in Afghanistan has fueled long-held fears in Islamabad of strategic encirclement. As Pakistan's conventional deterrent declines relative to India's heightened defense spending, Pakistan's nuclear deterrent becomes increasingly important. There are no guarantees that a conventional conflict between India and Pakistan will stay conventional. The greater the stresses endured by the Pakistani state, the less stable relations with India are likely to be. A Pakistan teetering on the brink of collapse is likely to act in unpredictable ways toward neighboring states and nonstate actors alike.

No greater threat faces the United States than nuclear material in the hands of terrorists. America's inability to deter groups like al Qaeda makes developing a comprehensive strategy to address this threat a vital national priority. The United States must invest in new ways of detecting loose nuclear material at home and abroad, but few have faith that we will be able to identify nuclear material crossing our borders if terrorists get hold of it. This heightens the importance of disrupting terrorist networks and limiting proliferation at its source.

The most direct ways to prevent terrorists from gaining control of nuclear material are to kill and capture terrorist leaders, limit the number of nuclear weapons states and stockpiles, and ensure the security of existing nuclear weapons arsenals. Each of these is a vital mission, but hard to achieve with any full measure of success. It was former Defence Secretary Donald Rumsfeld who asked in 2003 whether we are "capturing, killing or deterring and dissuading more terrorists every day" than are being recruited and deployed against us.[74] Five years later, we still do not have a concrete answer to the question. North Korean and Iranian pursuit of nuclear weapons demonstrates the complex geopolitics involved in trying to limit the number of nuclear weapons states. The United States has been more successful with reducing existing arsenals, but in Russia alone there still exist upwards of 10,000 warheads.

Classified plans to help secure nuclear stockpiles of partner states like Pakistan provide some assurances, but scenarios of state collapse could render such plans meaningless. One attempt to systematically look at collapse scenarios in Pakistan anticipated a requirement of one million troops to keep nuclear material from leaving the country, and concluded that, "it points to the critical importance of doing whatever is possible to prevent the collapse in the first place."[75] What, exactly, would be a plausible scenario in which terrorists could get hold of nuclear material? Determining this could help to determine which long-term demographic trends could be most worrisome. Stephen Cohen, one of the leading U.S. experts on Pakistan, has argued that all scenarios involving transmission of nuclear material to terrorists "lead back to the question of the army's integrity."[76] Possible transmission scenarios include:

- A hostile regime emerging through a coup, revolution, or election in which nuclear technology is transferred as a matter of policy to a terrorist group;
- Civil unrest that leads to divided command and control of the Pakistani military while a military faction proliferates nuclear material to terrorists;

- The continuing weakness of the state and armed forces to the extent that the security of nuclear stockpiles is in jeopardy, and material is stolen by a terrorist group.

Each of these nightmare scenarios demonstrates why U.S. policy has leaned heavily toward influencing state behaviour and building closer ties with Pakistan's armed forces since 9/11. During the 1990s, the Pakistani military operated largely outside of America's sphere of influence because of U.S. sanctions that were enacted after Pakistan's nuclear test. Of the 10-plus billion dollars in overt assistance provided by the U.S. Government to Pakistan since 9/11, over 60 percent has gone toward coalition support funds that reimburse the Pakistani military for its role in the war on terror.[77] This money, along with another billion-and-a-half in security assistance, has ensured the Pakistani military's cooperation and presence on the Afghan border, even if it has failed thus far to defuse the threat.

Recent tensions, however, between the United States and Pakistani militaries have risen to alarming levels and threatened to jeopardize the bilateral cooperation. A U.S. incursion into Pakistani territory by American Special Operations forces on September 3, 2008, prompted more than the standard rebuke from Pakistan. Pakistani military forces have since fired on U.S. helicopters and drones that have crossed or approached Pakistan's border with Afghanistan. It is impossible to predict whether these skirmishes will continue or escalate and how this might affect U.S.-Pakistan relations over the long run. The trend is worrisome, though, considering how vital the Pakistani military is to U.S. interests in Pakistan.

It is likely that some accord will be reached, and the United States will continue its close cooperation with the Pakistani armed forces. There is too much to lose for Washington not to resolve this crisis. The paradox, though, is that despite the Pakistan military's importance, the relationship has tended to frustrate other U.S. goals. America's ties to Pakistan's military have given the impression that the United States supports anti-democratic forces instead of the Pakistani people, provoking anti-American sentiment in the country.

The centerpiece of U.S. counterterrorism assistance in Pakistan today is a multiyear commitment to train Pakistani Special Forces and the paramilitary Frontier Corps in counterinsurgency doctrine and pour hundreds of millions of dollars in development money into the Tribal Areas. This attempt to win hearts and minds is the latest effort to work with local partners to create an environment in Pakistan unfavourable to al Qaeda and the Taliban. The difficulty, of course, is that our enemies are also seeking to shape this environment, and often times have proven more successful. While they may lack the resources we bring to the table, they have a veil of legitimacy from their cultural and religious kinship and anti-imperialist rhetoric that plays well to nationalist sentiment, even though their ideology is not nationalist itself. Pakistani public officials often speak of refusing to relinquish their sovereignty to foreign powers, but are too willing to accept the diminished sovereignty that comes from tolerating non-state actors like the Taliban and al Qaeda on their territory.

Ultimately, terrorists survive because of a lack of will to address the problem. As Henry Kissinger wrote one month after U.S. forces began their aerial bombing of Taliban-controlled Afghanistan in 2001:

The overwhelming majority of safe havens occur when a government closes its eyes because it agrees with at least some of the objectives of the terrorists. . . . Even ostensibly friendly countries that have been cooperating with the United States on general strategy...sometimes make a tacit bargain with terrorists so long as terrorist actions are not directed against the host government.[78]

The question that many in Washington have asked since 9/11—particularly after Musharraf's government cut a deal with militants in September 2006—is whether Pakistan has also made this tacit bargain.

Even now it is uncertain whether the civilian government in Islamabad sees the problem on the Afghan border as an insurgency that threatens the Pakistani state and is worth years of war and sacrifice to subdue. Many in Islamabad believe the violence against Pakistanis will dissipate once America eases its pressure and militant activity is again directed toward external targets. Many in Washington believe the United States has not yet found a true partner in Islamabad—civilian or military—willing and able to stand up to the Taliban and al Qaeda in both word and deed. This is why the United States has sought to take action into its own hands through unilateral military action, and why Pakistan has responded by firing on U.S. soldiers.

The United States may be the foremost power in the world, but the tools to protect ourselves from tomorrow's threats do not always lie in our hands. This is not necessarily cause for panic. Few countries in the world expect the freedom of action and ability to influence events that resides in American hands. It may be cause, however, for a reexamination of how America achieves its goals and the tools we need to succeed.

The United States must find ways to deepen its partnerships with foreign governments and militaries and key stakeholders in civil society to help shape an environment supportive of U.S. objectives over the long-term. No partnerships are perfect. The challenge will be improving those partnership where cooperation is inadequate but vital. As Defence Secretary Robert Gates has said, "the most important military component in the War on Terror is not the fighting we do ourselves, but how well we enable and empower our partners to defend and govern themselves,"[79] or as former Senate Armed Services Committee Chairman Sam Nunn is fond of saying, "We are in a race between cooperation and catastrophe."[80]

U.S. Policy Options

The first policy imperative of Pakistan's demographics is recognizing that as difficult a challenge Pakistan poses to U.S.

decisionmakers today, it will likely be magnified in a decade's time if action is not taken now. The 9/11 Commission said the United States should make a long-term commitment to Pakistan's future.

The underlying purpose of all action should be to mitigate the risk of nuclear material being transferred to terrorists over the long run. Although direct U.S. military action on Pakistani territory could prove necessary, the U.S. Government should do more shaping and influencing and less compulsion of friends, adversaries, and those in between.[81] Any direct action should be weighed against potential long-term consequences that could create conditions favourable to terrorist recruitment and broader conflict and instability in Pakistan. Are the targets of unilateral military strikes directly threatening to U.S. interests and lives? If not, the costs of stirring resentment in Pakistan may not be worth the immediate benefits of action.

U.S. shaping efforts should take the form of strengthening the Pakistani military's coherence and professionalism, promoting forces of political moderation, working to address divisions in Pakistani society, and building the capacity of government, military, and civil society actors. The U.S. Government already engages in much of this type of work, and yet the effect seems to be far less than the sum of the various parts.

The United States should recognize that its words and deeds can create incentives and disincentives for Pakistanis to work toward peace, stability, and moderation. It should be content to let politics play out in Islamabad without the shadow of U.S. interference. At the same time, the United States should quietly build deeper and more lasting relationships with all levels of the Pakistani military. The purpose of this engagement should go beyond general alliance maintenance and intelligence collection and seek to generate a common threat perception and set of shared goals.

America's visible presence in Pakistan should expand tremendously, but not along a security agenda. The Biden-Lugar bill for Pakistan gets many elements right: billions of dollars of aid for education and health over many years, greater accountability for security assistance, building a new relationship with the Pakistani people. It is a long-term prescription that is necessary for a counterinsurgency war that will take years to win.

Any long-term aid plan for Pakistan must include the following elements:

- **Massive new investments in teacher training:** America should become synonymous with quality education in Pakistan, not with the war on terror. The only way Pakistan will compete in the future is with strong public education.
- **Food, water, and energy assistance:** America should work closely with the Pakistani state and civil society to develop, fund, and implement a comprehensive programme to address resource shortages in Pakistan over the next 20 years. The United States has already provided shortterm food assistance to Pakistan, but longerterm programmes can be developed, particularly in the energy sector.
- **Trade assistance:** Even if political realities mean that the United States is not going to fully open its markets to Pakistani textiles, the United States should help Pakistan diversify and increase demand for its exports to lower its trade deficit. Provinces should have more say in the formulation of Pakistan's trade policy, and Punjab must become an engine of growth for all of Pakistan. Greater linkages must be built with China and India.

America's assistance to Pakistan should be closely tied to a strategic communication plan to help counter the ideology put forward by groups like the Taliban and al Qaeda. The initiative and implementation team should have an extensive local presence outside the American embassy, and be staffed by Pakistanis. There is added risk to operating country-wide at a time when anything associated with America could become a target, but working in Pakistan is a risky proposition, and the United States must be willing to bear more risk.

Ultimately, Pakistanis will need to make the sacrifices and tough decisions necessary to keep their country on a path toward peace and prosperity. The United States can exercise more patience, but at a certain point, Pakistan will need to demonstrate that it is committed to effectively reducing the militant threat on its western border. Tribal jirgas and a new, more balanced counterinsurgency strategy may prove to be the answer. The danger is that if these fail, the time horizons in Pakistan for U.S. decisionmakers are likely to get even shorter than they have been over the past 7 years.

Endnotes

1. Myron Weiner and Michael S. Teitelbaum, *Political Demography, Demographic Engineering*, New York: Berghahn Books, 2001, p. 2.
2. Richard Jackson and Neil Howe, The *Graying of the Great Powers: Demography and Geopolitics in the 21st Century*, Washington, DC: Center for Strategic and International Studies (CSIS), 2008, p. 12.
3. Weiner and Teitelbaum, p. 46.
4. Jackson and Howe, p. 19.
5. Tauseef Ahmed, "Population Sector in Pakistan: Current Demographic Situation Sectoral Problems and Issues and the Way Forward," *Multi-donor Support Unit*, No. 30610, October 2001.
6. Department of Economic and Social Affairs UN Secretariat Population Division, "Long-range Population Projections," proceedings of the UN Technical Working Group on Long-Range Population Projections, UN Headquarters, New York, June 30, 2003; the World Bank, "World Development Indicators Database: Country Data Profile," September 2008.

7. Population Reference Bureau, "Data by Geography," August 2004.
8. Zeba A. Sathar, "Fertility in Pakistan: Past, Present and Future," Workshop on Prospects for Fertility Decline in High Fertility Countries, Population Division, Department of Economic and Social Affairs, UN Secretariat, New York, July 9-11, 2001.
9. Durr-e-Nayab, "Demographic Dividend or Demographic Threat in Pakistan," *Pakistan Development Review*, Vol. 47, 2006, p.8.
10. *Ibid*, p. 7.
11. David Canning, "The Impact of Aging on Asian Development," Seminar on Aging Asia: A New Challenge for the Region, Kyoto, Japan, May 7, 2007, and Tokyo, Japan, May 8, 2007.
12. Durr-e-Nayab, p. 8.
13. Government of Pakistan, "The Demographic Dividend—Unleashing the Human Capital," Pakistan Development Forum, Jinnah Convention Centre, Islamabad, Pakistan, April 25-27, 2007.
14. Durr-e-Nayab, p. 9.
15. *Ibid.*
16. David Canning, p. 13.
17. Durr-e-Nayab, p. 15.
18. Ahmed, p. 3.
19. Durr-e-Nayab, p. 15.
20. Jackson and Howe, p. 141.
21. *Ibid*, p. 3.
22. Farhat Haq, "Rise of the MQM in Pakistan: Politics of Ethnic Mobilization," *Asian Survey*, Vol. 35, No. 11, November 1995.
23. Adeel Khan, Pakistan's Sindhi Ethnic Nationalism: Migration, Marginalization, and the Threat of Indianization," *Asian Survey*, Vol. 42, No. 2, March/April 2002.
24. Afghan Research and Evaluation Unit, "Afghans in Quetta: Settlements, Livelihoods, Support Networks and Cross-Border Linkages," January 2006.
25. Haris Gazdar, "A Review of Migration Issues in Pakistan," Migration, Development and Pro-Poor Policy Choices in Asia Conference, Dhaka, Bangladesh, June 22-24, 2003.

26. Devesh Kapur, "Remittances: The New Development Mantra?" G-24 Technical Group Meeting, Harvard University and Center for Global Development, August 25, 2003.
27. Aqila Khawaja, "Constraints and Challenges Arising from Demographic Transitions and Imbalances: Pakistan at the Crossroads," Network of Asia-Pacific Schools and Institutes of Public Administration and Governance Annual Conference, Beijing, China, December 5-7 2005.
28. *Ibid*, p. 560.
29. Jackson and Howe, p. 144.
30. UN Development Programme, *Beyond Scarcity: Power, Poverty and the Global Water Crisis*, New York: Palgrave Macmillan, 2006.
31. *Ibid*, p. 14.
32. Steve Morrison, *A Call for a U.S. Strategic Approach to the Global Food Crisis*, Washington, DC: CSIS, July 28, 2008, p.2
33. *Ibid.*, pp. 2-4.
34. Ather Maqsood Ahmed and Rehana Siddiqui, "Food Security in Pakistan: Can it be Achieved?" *The Pakistan Development Review*, Vol. 34, No. 4, part 2, Winter 1995, p. 723.
35. *Ibid.*, p. 724.
36. *Ibid.*
37. Ismail Quereshi, "Agriculture for Growth and Poverty Alleviation: Policies and Programs of the Government of Pakistan," Pakistan Development Forum: The Demographic Dividend- Unleashing the Human Capital, Jinnah Convention Centre, Islamabad, Pakistan, April 26, 2007.
38. *Ibid.*
39. Sohail Jehangir Malik, "Agriculture in Pakistan: Challenges and Prospects," Pakistan Development Forum: The Demographic Dividend- Unleashing the Human Capital, Jinnah Convention Centre, Islamabad, April 26, 2007.
40. Naser Faruqui, "Responding to the Water Crisis in Pakistan," *Water Resources Development*, Vol. 20, No. 2, June 2004, pp. 177-192.
41. *Ibid.*
42. Aaron T. Wolf, "Water and Human Security" *AVISO* 3, June 1999.

43. Government of Pakistan, "Pakistan: a Hydraulic Nation," Pakistan Development Forum: The Demographic Dividend-Unleashing the Human Capital, Jinnah Convention Centre, Islamabad, Pakistan, April 27, 2007.

44. Faruqui, p., 179.

45. Faruqui, p. 180.

46. Sandia National Laboratories, "Energy-Water Nexus Overview," available from *www.sandia.gov/energy-water/nexus_overview.htm*.

47. Mukhtar Ahmed, "Meeting Pakistan's Energy Needs," Woodrow Wilson International Center for Scholars Presentation, Washington, DC, June 2006.

48. *Ibid.*, p. 6.

49. *Ibid.*, p. 1.

50. Tauseef Ahmed, "Population Sector in Pakistan: Current Demographic Situation Sectoral Problems and Issues and the Way Forward," *Multi-donor Support Unit*, No. 30610, October 2001.

51. Rehana Siddiqui, "Energy and Economic Growth in Pakistan," *The Pakistan Development Review*, Vol. 43. No. 2, 2004, p. 175.

52. Robert G. Wirsing, *Baloch Nationalism and the Geopolitics of Energy Resources*, Carlisle, PA: Strategic Studies Institute, U.S. Army War College, 2008.

53. *Ibid.*, p. 8.

54. *Ibid.*

55. Michael Kugelman and Robert Hathaway, *Hard Sell: Attaining Pakistani Competitiveness in Global Trade*, Woodrow Wilson International Center for Scholars Asia Program, 2008.

56. Robert Looney, "Failed Economic Take-Offs and Terrorism: Conceptualizing a Proper Role for U.S. Assistance to Pakistan," *Asian Survey*, Vol. 44, No. 6, 2004, p. 781.57. *Ibid*, p. 771.

58. *Ibid*, p. 778.

59. Khawaja, p. 558.

60. UN Development Program, "Human Development Reports," available from *hdr.undp.org/en/statistics/*.

61. Shahid Javed Burki, "Pakistan's Economy: Challenges and Opportunities," CSIS presentation, Washington, DC, March 25, 2008.
62. Robert M. Hathway, ed., *Education Reform in Pakistan: Building for the Future*, Washington, DC: Woodrow Wilson International Center for Scholars, 2005.
63. *Ibid.*
64. Shahid Javed Burki, "Educating the Pakistani Masses: The World Needs to Help," Testimony before the Senate Committee on Foreign Relations on Combating Terrorism Through Education: The Near East and South Asian Experience, Washington, DC, April 13, 2005.
65. Robert Looney, "Failed Economic Take-Offs and Terrorism: Conceptualizing a Proper Role for U.S. Assistance to Pakistan," *Strategic Insights*, Vol. 2, No. 2, February 2003, p. 783.
66. Javed Hasan Aly, *Education in Pakistan: A White Paper,* National Education Policy Review Team, 2007, p. 39.
67. C. Christine Fair, "Militant Recruitment in Pakistan: Implications for al Qaeda and other Organizations," *Studies in Conflict and Terrorism*, Vol. 27, No. 6, 2004, p. 489-504.
68. A. H. Nayyar and Ahmad Salim, eds., *The Subtle Subversion:* The State of Curricula and Textbooks in Pakistan: Urdu, English, Social *Studies and Civics*, Islamabad, Pakistan: Sustainable Development Policy Institute, 2002, pp. 9-62, 70.
69. Graham E. Fuller, *The Youth Factor: The New Demographics of the Middle East and the Implications for U.S. Policy,* Washington, DC: The Brookings Institution, 2003.
70. Akbar S. Ahmed, *Resistance and Control in Pakistan*, New York: Routledge, 2004, p. 148.
71. Nicholas Schmidle, "Next-Gen Taliban," *New York Times Magazine*, January 6, 2008.
72. Fair, p. 490.
73. *Ibid.*
74. Donald Rumsfeld, "Rumsfeld's War-On-Terror Memo," *USA Today*, October 16, 2003.
75. Michael O'Hanlon, "Dealing with the Collapse of a Nuclear-Armed State: The Cases of North Korea and Pakistan,"

Princeton Project on National Security, Princeton, NJ: Princeton University, and Washington, DC: Woodrow Wilson School of Public and International Affairs, 2005.

76. Stephen Cohen, "Fractured Pakistan: Potential Failure of a Nuclear State," Fund for Peace Threat Convergence Conference, November/December 2006, p. 43.
77. Craig Cohen, "When $10 Billion Isn't Enough: Rethinking U.S. Strategy and Assistance to Pakistan," *The Washington Quarterly*, Spring 2007; Craig Cohen, *A Perilous Course: U.S. Strategy and Assistance to Pakistan*, Washington, DC: CSIS, 2007.
78. Henry Kissinger, "Where Do We Go from Here?" *The Washington Post*, November 6, 2001.
79. Robert Gates, "Landon Lecture," Kansas State University speech, Manhattan, Kansas, November 26, 2007.
80. Sam Nunn, "The Race Between Cooperation and Catastrophe," The American Academy in Berlin, June 12, 2008, available from *www.nti.org/c_press/speech_Nunn_Germany61208. pdf.*
81. Robert Gates, "Pre-Alfalfa Luncheon," CSIS speech, Washington, DC, January 26, 2008.

Chapter - 7

Imagining Alternative Ethnic Futures for Pakistan

—**Maya Chadda**

Introduction

Pakistan's growing crisis of governability is disturbing to policymakers across the world. A group of top U.S. experts on Pakistan conclude that "The United States cannot afford Pakistan to fail nor . . . ignore the extremists operating in Pakistan's tribal areas."[1] Why is Pakistan important to the United States, and what would make it more stable and democratic? The first question is less difficult to answer. Throughout the 19th century, Afghanistan was the cockpit of a titanic struggle that came to be called the Great Game. In a similar, but modern context, that role has now devolved to Pakistan. What makes its growing instability a particular cause of concern is that far from being a steadfast ally, capable of promoting and projecting U.S. interests in the region, internal developments in Pakistan are in danger of compounding threats to U.S. interests.

As a nuclear-armed, predominantly Muslim nation of 165 million, Pakistan is important to the United States in several ways. It is located adjacent to the oil rich Persian Gulf states and Central Asia; it borders on China and India, the two principal rising and rival states in Asia. As a result, Pakistan is strategically central to any attempt to prevent war and maintain peace in the region. All three states—India, Pakistan and China—are nuclear power states and historic rivals or allies. India is regarded as an adversarial state by both Pakistan and China, while China has built close security ties with Pakistan. At the same time, it is an Islamic nation with a declared interest in building a modern democracy, Pakistan can serve the United States both as a shield and as a sword; it can shield against expansion of radical Islam currently entrenched in the tribal areas between Pakistan and Afghanistan and act as a sword to eliminate their presence from the region. Should Pakistan become a successful democracy it would serve as an exemplar to the Islamic world in its own ideological battles.

However, Pakistan has had trouble establishing stability let alone a democracy over the past 65 years of its history. Military dictators have ruled the country close to 50 percent of the time.[2] The remaining years have witnessed populist leaders backed by cults of personality attempting to establish a party system but with little success given the way that the military has spread its tentacles throughout Pakistani society and permeated its institutions. Preoccupied with political survival, Pakistan's civilian leaders have paid little attention to reforms that could have prevented the state's slide towards failure. Even a cursory glance at Pakistan's history shows how it has lurched from one political crisis to another. The first decade of uncertain bureaucratic democracy ended in a military takeover in 1958 by General Ayub Khan, whose rule lasted for the next 10 years. Since then the army has, in a sense, never been out of power. Pakistan has slipped in and out of military rule three more times. It has been ruled directly by the military for 22 of the following 40 years, and even in the interregnums, civilian governments have only survived so long as the military acquiesced.

In retrospect, the 11 years of military rule from 1977 to 1988 under General Zia-ul-Haq was the beginning of Pakistan's steady slide into Islamization. Zia manoeuvred and revived a strategic alliance with the United States in which Pakistan agreed to serve as a conduit to Afghan Mujahedin fighting the Soviets forces. In return, the United States agreed to provide Pakistan with military assistance and turn a blind eye to its acquisition of nuclear technology, although officially the United States insisted on a policy of nonproliferation. In fact, the adventurous policy of cross border infiltration to foment rebellion in Indian Kashmir originated in the Zia years. To gain popular support for his illegitimate rule, Zia turned to the mullahs and Islamic leaders while Pakistan's civilian political leaders and parties were banned. The subsequent 10 years of uncertain democracy can be characterized as a period of diarchy, indirect rule by the military behind the façade of civilian rule and elections. Not one elected government was able to complete its term in office during those 10 years.[3]

The era of democratic experiment ended in war and a coup in 1999 by General Pervez Musharraf who dismissed Prime Minister Nawaz Sharif and arrested him on charges of corruption and mismanagement. These charge sheets were trotted out every time the Pakistani military decided to stage a coup and take over power and had become something of a template for takeover.

The year 2008 saw yet another transition from military rule to an uncertain coalition government, this time from General Musharraf to President Asif Ali Zardari and Prime Minister Yousuf Reza Gilani. But the new elected government must manage a Pakistan that is weakened by economic crisis, political discontent, and radical Islamists spreading violence and terror.[4]

Perspectives on Pakistan: The Debate

Pakistan began life with all the disadvantages of a newly born state. In sharp contrast to India, its lack of a preexisting state structure is the main reason why it succumbed repeatedly to military rule. The areas that were pulled together as Pakistan in 1947 had powerful local ethnic parties, such as the Awami League in East Pakistan and Khudai Khidmatgars in North West Frontier Provinces (NFWP). These parties consented to join Pakistan in the 1940s, largely because M. A. Jinnah, the leader of the Muslim League and founder of Pakistan, promised them self-rule and political autonomy. However, the *idea* of Pakistan as a homeland for the Muslims of the subcontinent had little space in it for ethnic and cultural diversity. The nationalist narrative of Pakistan was forged on the anvil of Islam.

The geographical anomaly of a single nation composed of two halves divided by the entire body of India only heightened the pressure on the early leaders to deemphasize its cultural and linguistic diversity. A geographically divided Pakistan could not become a democracy unless it accepted the possibility of electoral advantage going to its more populous eastern wing. This proved too much for the West Pakistan based military-feudal elite to swallow.

Despite the turbulence caused by repeated military coups, the surgical dissection of the country in 1971, and the temporary loss of U.S. interest in South Asia after the Vietnam War, every Pakistani government adhered firmly to two principles of foreign policy: strong security ties with the United States, and an enduring conflict with India. Conflict with India was inevitable because of the way the partition had occurred in 1947, unleashing a communal holocaust and leading to war between India and Pakistan over Kashmir. Each of these relationships contributed to the ascent of the military to power in Pakistan. A hostile India justified putting the armed forces in charge; alliance with the United States helped Pakistan defend against the larger India. But this strategy came at huge domestic costs because swollen military budgets preempted expenditure on social development, particularly education and health.

It is against this complex intertwining of domestic and international forces that we need to imagine an alternative ethnic future for Pakistan. What would make it more stable? What would lead to a consolidation of democracy in Pakistan? There is no consensus among observers in the United States on how this can be achieved. One view advocates a firm adherence to the conventional road of free and fair elections. It believes that this will guarantee the inclusion of all sections of society and facilitate the emergence of a stable polity. A second view is that political institutions need to be strengthened first, through a reform of the political parties as well as the legal and constitutional framework of the state. The conclusion is that only then will free elections yield the desired results. More communityoriented observers stress education, health, and transparency and consider advances in these to be a necessary precondition for the emergence of a stable democracy. There is thus no clear consensus on where to start and how to proceed.[5]

In the United States this debate is understandably focused on rolling back the advancing Taliban groups that are now entrenched in the Federally Administered Tribal Areas (FATA) and the North West Frontier Province (NWFP).[6] Under U.S. pressure, Pakistan's government abandoned the strategy of negotiating with some factions of the Taliban, provided they gave up arms and accepted the suzerainty of the government in Islamabad. General Musharraf had strongly supported the deal and the later leader of the opposition in Pakistan's Parliament, Nawaz Sharif and Prime Minister Gilani also thought it to be a prudent strategy. The idea was to use a carrot and stick approach to divide the Taliban and regain control of the FATA. General Musharraf was convinced that the FATA areas had to be incorporated into Pakistan, but this meant changing its current status and turning it into another province of Pakistan. As a result, the deal was to be an interim arrangement.

But the idea of a deal with the radical Taliban was deeply disturbing to the U.S. Government. It smacked of appeasement and weakness; it also institutionalized the territorial gains made by the Taliban. This was not acceptable to the United States largely because it undermined its goal of weakening the Afghan Taliban and stabilizing the Afghan government under a non-Taliban rule.

Many in the United States argued that the Taliban had a larger agenda and allowing them to gain legitimacy would threaten

Pakistan while undermining in the interim U.S. efforts at strengthening the Karzai government in Kabul. The short-term U.S. objective was to help Pakistan's military eliminate the Taliban, but strengthening the Pakistani army created serious problems for all future civilian governments. And many in the policy circles close to both the Clinton and Bush administrations had argued that in the ultimate analysis, the only effective answer to Islamic radicalization was a stable and democratic Pakistan. But how could the United States promote democracy while supporting the military in Pakistan? The U.S. strategy of backing the military had produced only aborted democracy in Pakistan. It was unlikely to succeed this time around. Neither the objectives of democracy nor the narrow focus on the U.S. war on terrorism tells us how Pakistan can build a stable state as a first step towards democracy.

This chapter seeks to explore ways in which this objective can be achieved despite the political constraints under which any Pakistani government has to operate. I argue that democracy is not a panacea for instability, at least not in the short run. Democratic competition can exacerbate conflict, and, while democracy ought to be the goal, ethnic reconciliation and conflict management capacity are more important as first steps towards that goal. This interim course of action does not need a full-fledged democracy to be in place. Pakistan has possessed a partial and rather unsuccessful federal system since at least 1973, although its record of accommodating its nationalities has been dismal. Integrating these in an enduring way is an existential imperative for a stable Pakistan. If the United States is interested in Pakistan's stability, then it needs to help Pakistan find a formula to forge a new ethnic bargain that will revive its federal mandate.

Pakistan is not alone in having to balance ethnic and regional influences against the need to unify; nor is it the only country in South Asia to fear disintegration because of ethnic overlap, religious fundamentalism, civil strife, and disputed borders. In all these respects, countries in South Asia share Pakistan's problems: One is an ethnocracy (Sri Lanka), two are partial and episodic democracies (Bangladesh, Nepal), while one is a fully-fledged but still flawed democracy (India). South Asia provides a valuable context to imagine an alternative future for Pakistan. This context suggests that, short of a fully functioning and vibrant democracy, a revitalized power-sharing agreement between the central Pakistani state and its parts can be a viable path to stability.

Nature of Violence and Conflict in Pakistan

Ethnically and linguistically, Pakistan, like neighbouring India, is one of the most heterogeneous countries in the world. At its inception, Pakistan was made up of five large distinctive nationalities, Bengalis in East Pakistan and Sindhis, Punjabis, Pushtuns, and Balochis in West Pakistan. There was a sixth large group, but it had no territorial base: the Muhajirs, who had migrated to the newly created Pakistan and settled largely in Punjab and Sindh. Leaders of Pakistan presumed that Islam would hold the fragile union together, but support for the idea of a separate nation of Pakistan had been tepid among the Baloch, Pushtuns, Sindhis, and even to some extent among the Punajbi Muslims, until weeks before the partition. These ethnic communities had agreed to join Pakistan on the promise of a large degree of autonomy and selfrule. But Pakistan's post-independence history belied these hopes, and produced instead a serious imbalance between ethnic nationalities and the central state.

Pakistani leaders shared a common fear in the early years of independence that granting concessions to ethnic nationalities might spiral into separatism. Exhorting his countrymen, M. A. Jinnah, Pakistan's founding father had said, "You have carved out a territory, a vast territory. It is all [yours]; it does not belong to a Punjabi, or a Sindhi, or a Pathan or a Bengali. It is all yours." Therefore, "If you want to build yourself up into a nation," he said, "for God's sake, give up this provincialism."[7] And although India was committed to creating a federal structure from the very beginning, India's first Prime Minster, Jawaharlal Nehru, had voiced strong misgivings about provincialism and railed against it in the early 1950s.

Nehru and Jinnah had good reason to fear provincialism. No state can be formed without some concentration of power at the center, and the power has to be exercised most decisively at the very beginning of nation-building.[8] Progressive consolidation, however, requires an institutionalization of the means to circulate and share power. Democracy, or at the very least, negotiated power sharing are the only ways to ensure that this takes place.

Neither development had occurred in Pakistan. India offers an interesting contrast. "In India, federalism was the mechanism to accommodate great linguistic heterogeneity, creating multiple identities. But the elite of Pakistan viewed regional and linguistic identities as inherently dangerous and as undermining the 'nation project'. The adoption of Urdu as the state language was an indication of interregional identity projected by the center."[9]

While Pakistani leaders feared power-sharing, the denial of autonomy was precisely what led to the breakup of East from West Pakistan in 1971. The loss of Bengali-dominated East Pakistan, however, made accommodation with ethnic identities more, and not less, urgent. The Bengali majority in undivided Pakistan had counterbalanced the Punjabi majority in West Pakistan to give the other ethnic groups the political space to assert themselves. The secession of East Pakistan therefore triggered powerful movements for provincial autonomy for a Sindhu Desh, an independent Balochistan, a NWFP tied to Afghanistan, and even a Mohajir state that aspired to turn Karachi into another Singapore. What complicated matters even more was the injection of Islamist ideology into FATA and northern Balochistan after 1990.

Why did these grievances eventually morph into a violent movement against the state of Pakistan? In addition to denial of autonomy, two other factors were responsible: the preponderance of Punjabis, not only in numbers but also in wealth and power within the army; and the repeated military takeovers of Pakistan's government after 1971. Together these three conditions destroyed the political channels that might have established a new more equitable balance between Pakistan's ethnic nationalities and its central state.

The Destiny of Constitutional Provisions to Provincial Autonomy

If Pakistan did not become a democracy, it was not for want of trying. Between 1947 and 1956, there had been no constitutional representation in the newly created Pakistan. The constitutional crisis that developed during the existence of the first Constituent Assembly strengthened the role of central institutions—the bureaucracy and army—at the expense of regional parties in the provinces. The 1956 constitution and the governments it created were short-lived and gave way to political chaos. Acutely conscious of its weakness in comparison to the well-led larger India, Pakistan's elite opted for military rule to jettison democracy. The unsuccessful conclusion of the first war with India in 1948 had already frozen the Indo-Pakistan frontier into a hostile border. By 1956, the Cold War was jelling into military alliances led by the United States and Soviet Union. It is in this context that General Ayub Khan decided to assume control in 1958, suspend the constitution, and link the course of Pakistan's foreign policy to U.S. containment objectives.

President Ayub ruled the country with an iron fist and ignored the demands of Pakistan's ethnic nationalities. Instead, he strengthened the role of the bureaucracy and military within a new constitutional set up that transformed Pakistan into two administrative units, West and East Pakistan. The purpose was to balance the Bengali-dominated East Pakistan and to deny the latter power proportionate to its numerical majority. But it also had the unfortunate side effect of forcing the merger of all ethnic nationalities in the west—the Baloch, Pushtun, Punjabi, Sindhi and Muhajirs—into a single unit.[10]

The single unit plan (meant to establish a parity between the two halves of Pakistan) was the starting point of all of Pakistan's subsequent problems, and indeed of its drift towards Islamism in the 1980s. It sowed the seeds of secession in East Pakistan. It also effectively disenfranchised the Sindhis, the Balochis, and to a lesser extent the Pushtuns. The years between 1969 and 1972 were chaotic and violent. Ayub Khan's fall from power led to the first authentic elections, but these resulted in a majority for the Awami League party of East Pakistan. Unable to tolerate the prospect of an East Bengali prime minister, General Yahya Khan staged a preemptive coup that settled the fate of Pakistan's democracy for the second time since its birth. What followed was a civil war, intervention by the Indian army, Pakistan's defeat, and the secession of East from West Pakistan. Since the military had been discredited, political parties, particularly the Pakistan People's Party (PPP), emerged as the alternative. Prime Minister Bhutto gave Pakistan its third constitution. This constitution has been in abeyance for much of the last 3 decades, but it subsequently became the basis for constitutional modifications. These occurred largely in response to the shifts in the balance of power between the military and civilian leadership in Pakistan. The 1973 constitution repealed the One-Unit Plan (returning West Pakistan to the original four provinces and tribal areas) and put in place provisions for regional autonomy within a federal state. Its adoption was tantamount to an admission that ethnic accommodation was an existential imperative for Pakistan.

The 1973 constitution, framed in the aftermath of Bangladesh's secession, formally restored the principle of federalism, redefining the term as maximum provincial autonomy. The residual powers were vested in the Provincial Assemblies, and for the first time a bicameral legislature was elected. The Senate was elected for 4 years on a basis of regional parity. The provinces, Punjab, Sindh, NWFP and Balochistan were to elect 14 members each for 4 years and half of the members retired after 2 years.[11] The 1973 constitution contained two lists: Federal and Concurrent. The Federal list comprised two parts. Part I contained items over which only the Parliament could legislate, encompassing 67 subjects. Either the Federal and Provincial governments could legislate over the items in the Concurrent List; however, in case of conflict over the exercise of power, the central government's right prevailed (article 143).

The 1973 constitution created a federal structure, but it did so in the absence of any genuine understanding, or indeed respect, for the federal principle. While it devolved a large share of legislative power on the provinces, Prime Minister Bhutto remained deeply reluctant to devolve power to the provinces. This became evident in his treatment of Baloch nationalism.

The weaknesses of the One-Unit Plan had been exposed by events in Balochistan even before the secession of Bangladesh. In an effort to curb the growing secessionism there, General Yahya Khan had granted Balochistan the status of a separate province in 1970. But in the aftermath of Pakistan's breakup, fearful that Balochistan would go the same way, Zulfiqar Ali Bhutto, operating as interim President under an interim constitution, dissolved Balochistan's coalition government led by Attaullah Mengal on February 15, 1973. When the National Awami Party (NAP) and Jammat-Ulemma–Islami (JUI) coalition in the NWFP resigned in protest against Bhutto's arbitrary action, he drew no lessons from it and instead banned the NAP in February 1975 and arrested its leaders under charges of conspiring against the state. They remained behind bars until 1977. As a result, throughout the Bhutto period there was no effective opposition in the National Assembly.

Federalism received a body blow when General Zia-ul-Haq engineered another military coup in 1977. For the following 8 years, the 1973 constitution went into abeyance, and federalism came to be substituted by a party-less democracy which slid rapidly toward Islamization of Pakistan's state and society. The accidental death of General Zia brought another 10 years of uncertain and unstable democracy to Pakistan, but no elected governments finished their terms during those 10 years. The military dismissed ruling coalitions led by Benazir Bhutto and Nawaz Sharif two times each on the standard charges of mismanagement and the endangerment of national security. It was clear that Pakistan's constitution would operate in breach and federalism would disintegrate under the force of struggle among Pakistan's political elites. Indeed this period witnessed the crass manipulation of ethnic movements through the tactics of divide and rule, cooptation, and bribery. Ethnic parties participated in the four coalition governments, but the fear of military intervention exacerbated uncertainty and let loose the very worst features of democracy in Pakistan: alliances for pecuniary and political purposes that were devoid of a larger purpose.

The only power-sharing experiments worth noting during the Zia and then the Musharraf governments were the devolution plans each had introduced. These plans came at different times in Pakistan's history but were similar in motivation, general outline, and purpose. The International Crisis Group (ICG) report on these plans points out that "The primary motivations for Zia to create local bodies was to legitimize the military government, broaden its support base beyond the military, and use the newly created and pliable local elite to undermine its political opponents."[12] In essence, the local bodies provided the "civilian base of his military government, supporting it in return for economic and political benefits."[13] Gradually, these local governments became a vast mechanism for extending state patronage to promilitary politicians, providing the military government with ample scope for staging favourable, nonpartisan elections. In due course, the new local elites formed the core of Zia's rubber stamp parliament, elected in nonparty national elections in 1985. But these local bodies could not assuage popular demands for participation or bestow any lasting legitimacy on the military government.[14]

The report goes on to the say that "Devolution, in fact, has proved little more than a cover for further centralized control over the lower levels of government" in Musharraf's plan. The ICG report also points out:

Despite the rhetoric from Islamabad of empowerment . . local governments have only nominal powers. Devolution from the centre directly to the local levels, moreover, negates the normal concept of decentralisation since Pakistan's principal federal units, its four provinces, have been bypassed. The misuse of local government officials during the April 2002 presidential referendum and the October 2002 general elections has left little doubt that these governments were primarily instituted to create a pliant political elite that could help root the military's power in local politics and displace its traditional civilian adversaries.[15]

Time and again, Pakistan was denied democracy, which could have welded the country into a coherent nation-state. It could not settle on a legal political framework that could have channeled protests and integrated its diversity into a coherent whole. Still, the absence of a legal political framework alone cannot explain why Pakistan's ethnic nationalities turned to violence and separatism. For that we need to briefly sketch a short profile of their key grievances.

Pakistan's Ethno-nationalities: Plotics of Discord

Sindh

Sindhi separatism can be traced all the way back to the group's marginalization in the creation of Pakistan in the 1940s and to the demographic changes that followed Partition. Sindhi is not only a distinctly different language from Punjabi, Urdu, and Pushtu, it has a rich and long literary tradition. Partitioned Sindh, not unlike partitioned Bengal in 1947, was divided between the more modern, urban, and prosperous Hindu minority and the feudally dominated Muslim peasantry in rural Sindh. During the partition most of the Hindus migrated to India while Urdu-speaking Muslim refugees (Muhajirin) settled in Sindh, particularly Karachi. These migrants were better educated and soon entrenched themselves among Pakistan's ruling elites. In contrast, Muslim Sindhis lagged far behind with only 10 percent levels of literacy compared to 70 percent among the new comers. Wright comments:

This handicap might have been compensated politically as Sindhis still had a two-thirds majority in the province as a whole had it not been for the exigencies of national politics in the new country. During the first republic of Pakistan (1947-58) and certainly until the assassination of Prime Minister Liaquat Ali Khan in 1951, politics was dominated by the refugee leaders from northern India and Bombay, but they lacked constituencies and were consequently reluctant to hold national elections."[16]

The first cause of friction between the government of Sindh and the central government was the choice of Karachi as the national capital. The city of Karachi, which the Sindhis claimed always to have belonged to Sindh, was "demarcated and placed in 1948 under central administrative control."[17] Later Sindh was merged into West Pakistan as a result of the Two-Unit plan.

In 1960, President Ayub Khan moved the national capital to Islamabad leaving Sindh without the national capital or any capital whatsoever. Ironically, this also marginalized the Muhajirins (refugees) who had migrated to Pakistan in 1947 and settled in Karachi in large numbers. In the early 1950s, attempts to impose Urdu as the national language led to rioting and demonstrations in Sindh, but the government adamantly replaced Sindhi as a medium of instruction with Urdu. The espousal of Urdu as the national language was implicit in the Two-Unit plan and the centralization under the military government in Pakistan. The imposition of Urdu on East Pakistan led to similar rioting and violence and forced the government to withdraw the directive that declared Urdu's exclusive national status.

In 1970, Ayub Khan's successor, Yayha Khan, returned Karachi to Sindh, but by then the conflict between the Sindhis and Muhajirins had become institutionalized. It was also a harbinger of the extreme communal violence that was to tear the city apart in the 1980s and 1990s. Intra-ethnic tensions in Sindh were caused largely by the Ayub government's policy to settle Punjabi officers in Sindh through land grants; especially irrigated land along the Indus. Pakistan had failed to carry out land reform in a system where the distribution of ownership was extremely uneven and dominated by an oppressive feudal system. This was particularly so in Sindh.[18]

Political interference by central governments as well as the manipulation and repression of local leaders added further fuel to popular discontent, but the fate of Sindhi nationality was no different from that of the other nationalities in Pakistan. When Sindh's first chief minister, Mohammed Ayub Khushro, "opposed the establishment of Karachi as the federal capital in 1948, he was dismissed by his rival and fellow Sindhi, the provincial governor, G. H. Hidayatullah, on grounds of maladministration and corruption, although Khushro still had the support of a majority in the assembly." This was only the first in a series of interventions that culminated in a bitter fight over the imposition of the Two-Unit plan. Political manipulation forced Sindh to accept the plan but "Sindhis were [left] without an adequate voice to represent their aspirations and concerns."[19]

"This process was repeated under General Ziaul- Haq's martial law regime (1977-85), but even the second Sindhi Prime Minister, Mohammed Khan Junejo (1985-88), encountered major dissidence in his home province."[20] In early 1970s, Prime Minister Z. A. Bhutto had tried to address the Sindhi grievances but he could do little, "perhaps because," explains Wright, "he did not dare antagonize either the army or Punjabi voters on whom he relied for continuance in power." [21]

While Sindh has not declared open rebellion against the Pakistani state, tensions continue to fester even today, and resentments have accumulated that flare up frequently in the form of violent confrontation between Sindhis and Muhajirs in Karachi. Indeed Karachi, a huge city and the hub of commerce and trade in Pakistan, presents a special case of inter-ethnic conflict. Economic factors, demographic pressures, and militant Islam have turned Karachi into one of the most unsafe cities in the world.

Muhajirs

The motivating cause behind the mobilization of the Muhajirins was different. It occurred not because of discrimination or lack of representation, but because of the Muhajir leaders' gradual loss of status and influence among the Pakistan's ruling circles. This became especially pronounced after 1971 and the secession of Bangladesh. In 1979, in response to events in Karachi, the Muhajirs founded the All Pakistan Muhajir Students Organization (APMSO) to compete with other ethnic student groups, particularly the Jama'at-i-Islami youth group, the Jamiat-e-Tulaba, on campus. In March 1984, when President Zia–ul-Haq banned all student organizations, Altaf Hussain, then the head of the APMSO, transformed his student organization and founded the Muhajir Qaumi Mahaz (MQM), a party that dominated Karachi throughout the 1980s and 1990s. In August 1986, it issued a charter of demands that included full representation in provincial as well as federal government departments on the basis of population; the grant of voting rights to the real Sindhis and Muhajirs, while non-Sindhis (including non-domiciled and non-Muhajirs) were to be given only a business permit to operate in Sindh; the setting of quotas and reservations for Muhajir students; a ban on outsiders buying property in Sindh; an extension of citizenship rights to Bengali Muslims (Biharis) stranded in Bangladesh, the confinement of Afghan refugees in their camps and the nationalization of bus services owned by "Pathans" (Pushtuns). Most importantly, the MQM demanded that its people be recognized as the fifth nationality of Pakistan, along with Punjabis, Sindhis, Pathans and the Baloch.[22]

Baloch and Pushtun Ethnic Demands

While disaffection simmered in Sindh, it was in Balochistan that ethnic resentments burst into flames. After the secession of East Pakistan, Baloch separatism became the most dangerous challenge to the security and authority of governments in Islamabad. The previous discussion has already explained why tensions exploded in 1973 and how the government of Z. A. Bhutto crushed the Baloch revolt. But Baloch nationalism has refused to die out. There are at least three reasons why the Baloch demand a separate state: changing demographic distribution in the province, the effects of socio-economic modernization on the traditional Baloch life style, and the enduring struggle for power in Pakistan.

The Baloch are a small minority in a large province, accounting for only 2 percent and then 5 percent of Pakistan's population before and after the secession of East Pakistan. Compared to Punjab and Sindh, the Baloch province is grossly underdeveloped. The Baloch have migrated to other provinces in search of jobs. This out-migration has reduced their number in their ethnic homeland. At the same time, a huge number of Punjabis, Pathans, and Sindhis have migrated to Balochistan in search of economic opportunities. The migration from Afghanistan increased markedly after the 1978 Marxist coup in Kabul and the Soviet invasion of that country in 1979. Rajat Ganguly writes, "The consequences of demographic transition in Balochistan have been severe for the continued cultural integrity of the Baloch and their political control of Balochistan."[23]

The demographic transformation brought modernization, which has in turn undermined the Sardari system (indirect rule through tribal chieftains) that the British had put in place. But the weakening of tribal traditions has not advanced to the point where the Baloch region can be smoothly incorporated into Pakistan. "The ruling elite in Pakistan", writes Ganguly, "in their attempt to build a strong, centralized state made it imperative to break down the power of tribal chieftains as part of a larger effort to merge Baloch identity into an all embracing Pakistani identity."[24] The central government built roads to make Balochistan more accessible and increased the number of army retirees (largely from Punjab and NWFP) that were settled by land grants in Balochistan. This caused friction among the native Baloch and the new migrants. The tribal Sardars revolted, and the educated and better-traveled among them (for example, Khair Bax Marri and Mengal, who were both heads of their respective tribes) opposed Balochistan's incorporation into Pakistan.

The Pushtuns[25]

Pakistan's frontier provinces are populated by the Pushtun who live on both sides of the Durand line marking the border with Afghanistan and have harboured irredentist aspirations from time to time. Left alone to live according to their customs, the ethnic tribes glorify,

"independence, battle and personal bravery, and deeply tribal code of honour (Pukhtunwali) whose three cardinal tenets are revenge, sanctuary, and hospitality. These tribes have ethnic connections with the tribes in Afghanistan; many tribal families in fact live on both sides of the Durand line. Movement across the border has been free and unhindered for hundreds of years."[26]

Pakistan's governments have followed a two-pronged approach to the tribal regions: cooptation of vocal and powerful ethnic elites, and neglect of the rest or repression when a recalcitrant ethnic leader refused to toe the line. On the whole, cooptation of Pathan nationalism has worked far better than that of the Baloch largely because the Pushtun rose to become officers and were recruited into Pakistan's armed forces in large numbers. This deferential approach to the tribal areas, as opposed to the other provinces of Pakistan, had tacit support from the United States. Rajat Ganguly writes, "Pakistan's improved military and financial capability as a result of joining the western bloc also allowed the central regime to implement the carrot-and-stick policy more effectively."[27]

Had Afghanistan not been invaded by the Soviet Union in 1979 and by the United States in 2001, Pushtun ethno-nationalism would have been no different from the Baloch or Sindhi variants. Indeed, because the Pushtuns were heavily integrated into the framework of the Pakistani state, they would probably have been assimilated into a new system of power sharing. But 1979 changed all that.

The Soviet invasion completed the split that had existed since the end of the 19th century between the urban, modernized, and largely Russian-influenced Afghani elite in Kabul and the deeply conservative and traditional tribesmen in the countryside. War against the godless heathen radicalized the latter. This was actively encouraged by both the United States and Pakistan for their own self-serving purposes. The ground was then laid for the conversion of a simple ethnic movement into a complex insurgency that fused religion and nationalism. This conversion took place in three stages. During the Afghan war, Pakistan's tribal areas and the NWFP became sanctuaries for the oppressed and insurgent Afghans alike. There was no conflict between Pushtun ethno-nationalism and the Pakistani state during this period. This began only after the United States declared a war on the Taliban regime in Afghanistan in 2001, breaking the links between Afghan nationalism and Pakistan to the severe detriment of Islamabad's relations with its own Pushtun population.

The third stage followed naturally from the previous break between countryside and capital in Afghanistan. As the Taliban and large numbers of nationalist Pushtuns took shelter in Federally Administered Tribal Areas (FATA), Pakistan was forced to join the war against the Taliban. This turned the complex brew of nationalism and religion, which had so far targeted the United States and the North Atlantic Treaty Organization (NATO), inwards and against the state of Pakistan. The sporadic truce that had prevailed between the government of Pakistan and the radicalized Pushtuns ended when the Pakistan army attacked the Islamic radicals hiding in the Lal Masjid in Islamabad in July 2007.

The current conflict in Pakistan is often characterized as a conflict of modernity with retrograde Islam, or it is described as a clash between more radical and deeply conservative Islam. Only tangentially have the commentators acknowledged the role of Pushtun ethno-nationalism in it. The current conflict is all of the above and more. It is within this context that Pakistan has to defend its northwestern borders and prevent violence from spreading to its provinces and cities. But it has not been able to do this. Islamist extremism has gained a strong base in the NWFP ever since the Islamist alliance, the Muttahida Majlis-e-Amal (MMA), secured an absolute majority in the Provincial Assembly elections of October 2002. Islamic parties, which never received more than 2 percent of the vote in any previous elections, are now commanding double digit support. In fact during the past 5 years, the FATA and the NWFP have become the primary arena of struggle between militant Islam and Pakistan's modern state.

The erosion of central control is reflected in the NWFP's slide from a fairly well-integrated province into a rebellious region. The Pakistani state previously had a strong presence in and a widely accepted powersharing arrangement with NFWP.[28] Its situation was not even remotely comparable to the lightly and indirectly governed FATA. But the consolidation of the Taliban/ al Qaeda axis in these regions has forced a difficult choice on Pakistan: attempts to incorporate the tribal regions might mean a long war in the region and against its own people; failure to do so will strengthen radical Islam and spell an end to the moderate Muslim state that Pakistan can become. Aware of these dangers and pressured into action by the United States, General Musharraf temporized by ordering offensives in 2006-07 in Balochistan, and in the FATA in 2007-08. In Balochistan the free rein given to the army drove the insurgency underground.[29]

In FATA, Musharraf's strategy was more ambivalent. This was because the vast majority in Pakistan was uncomfortable with the idea of deploying the army against its own people in these regions. The average Pakistani regards the Taliban as misguided youth that deserve understanding and sensitive handling. Bin Laden is popular and the United States, particularly since the military strikes, increasingly unpopular.[30] The subsequent reign of terror against civilians tilted the public against the Taliban, but the raids have cancelled this out.

The deadly connection between Pakistan's home grown Islamic radicals and the war in Afghanistan became fully visible in the aftermath of the July 2007 Lal Masjid episode. The Tehreek-e-Nafaz-e-Shariate-Mohammadi (TNSM) or the Movement for the Enforcement of Islamic Laws which operates in the Swat district of NWFP, broke the peace agreement it had signed with the provincial government on May 22, 2007,[31] and declared a jihad against those responsible for the military assault on the mosque. What followed came close to civil war. Violence is no longer confined to the tribal belt between Pakistan and Afghanistan.

The battle between the Security Forces (SFs) and Tehrik-e-Taliban Pakistan (TTP) intensified in 2009. According to official data, 1,400 militants were killed in a military offensive that commenced on April 26, 2009, even while close to 3.8 million people were fleeing their homes in search for safety and succor. The operations were initially confined to Lower Dir, Buner, and Swat Districts of the NWFP. These were spread to the rest of the FATA and NFWP. While the SFs have stepped up their operations, the TTP has responded with a welter of attacks across Pakistan's urban areas and elsewhere.[32]

What is more, it has spread to Punjab, the heartland of Pakistan. During the first half of 2009, 155 persons, including 92 civilians and 51 SF personnel, were killed in 104 terrorism-related incidents in Punjab. Only 12 terrorists, including nine suicide bombers, were killed. This may be because the Taliban/al Qaeda network is securing an upper hand in areas beyond the tribal belt. Even the nation's capital, Islamabad, Punjab's provincial capital, Lahore, and the garrison town of Rawalpindi have not escaped terrorist attacks. Out of the 104 incidents recorded in the first half of 2009, nine were reported from Islamabad and 18 from Lahore.

Perhaps the most dangerous outcome of the conflict in FATA has been the rapid replacement of old, conservative ethnic Pushtun tribal leaders with younger and more radical aspirants seeking to lead a Jihad against all its enemies—the Pakistan government and security forces, the Americans, NATO, and the pro-Pakistan co-opted tribal leaders. The radical leaders and their followers quarrel among themselves but neither the United States nor Pakistan has been able to benefit from it. As India learned to its consternation in Kashmir, divisions among the militants only prolong the conflict.

In summary, the convergence of ethnicity, Islamist fundamentalism, Pashtun nationalism, and a hatred of the West fostered by al Qaeda, has created a qualitatively different situation from anything that Pakistan has ever faced before in FATA, parts of Balochistan, and a widening swathe of the NWFP. It is a challenge for which the feeble and still very young state of Pakistan has no effective response. The solution for Pakistan lies in separating radical Islam from nationalism, particularly, ethnic nationalism. South Asia's experience suggests that conflicts based on religious ideology are more difficult to resolve because, unlike culturally defined linguistic and regional identities, they are not amenable to resolution through power-sharing and cultural accommodation. Ethnic movements do not usually extend beyond the boundaries of the ethnic homeland. It may be easier to negotiate and settle with the Baloch and the Sindhis by granting them a large measure of provincial autonomy than to do so with the multiplicity of Islamic radical groups operating in FATA, who are determined to capture Pakistan and establish an Islamic Caliphate from Kabul to Srinagar.

None of the policy prescriptions for Pakistan being mooted in the United States come to grips with the sheer complexity of the challenges they face. Pakistan's elites share the deep fear of the Islamic radicals with the United States but the single focus of the United States on them has made it convenient for Pakistan to ignore the problem of ethnic self-assertion. This has given the military and civilian governments the pretext to postpone the search for power sharing strategies that might stabilize Pakistan.

Implication for Pakistan

Seth Jones observed in a well-argued RAND report that, "Every successful insurgency in Afghanistan since 1979 has enjoyed a sanctuary in Pakistan and assistance from individuals within the Pakistan government, such as the Frontier Corps and the Inter-Services Intelligence Directorate (ISI)."[33] To restore peace to Afghanistan, Pakistan must be stabilized and made free of insurgency and the support base it offers to the Afghan Taliban. Failure to do so will cripple long-term efforts to stabilize Pakistan, rebuild Afghanistan, and might even jeopardize India.

The United States has been following a three-pronged strategy. The first is to press Pakistan to use coercion against the extremists while enlisting the support of tribal chieftains and local leaders against the insurgents with economic and political incentives. The second prong is to improve governance by building schools, clinics, roads, and other social projects, with the aim of infusing confidence, increasing the visibility of the central government, and gradually integrating these areas into the mainstream of Pakistan. The third prong is to secure a stable government and a dependable leader in Islamabad.

But this policy has not worked. Instead, the replacement of Musharraf by a far weaker, albeit democratically elected, government under President Asif ali Zardari and Prime Minister Yousuf Reza Gilani has compelled the Bush and Obama administrations to step up U.S. military raids in the border areas.

While strengthening of democracy must wait for the success in the war, the two are closely intertwined. Many therefore wonder whether the military option will open the space for a stable democracy and whether the small and weak opening to democracy after Musharraf can be transformed into a vigorous and stable democracy in Pakistan. The current debate among students of Pakistan is polarized between optimists who advocate a rapid advance to fullyfledged democracy and pessimists who are willing to settle for an authoritarian regime because they fear chaos more than they desire democracy.

This debate has discounted the possibility of finding an interim solution that looks for a way to expand the number of stakeholders in the democratic process. One way to do this is to reapportion political power between the state and its diverse and so far largely unintegrated ethnic communities. This requires understanding democracy in a way that is different from the way in which it traditionally has been understood in the west, i.e, as a relationship between the state and individuals in society. That definition is characteristic of unitary states and precludes the possibility of layered sovereignty and plural citizenship. In South Asia, sovereignty has always been layered, and the individual's connection to the state has been mediated through kinship groups and regional and cultural identities that have a prior claim.[34]

Ethnic identities can be efficient, if not ideal, building blocs for a liberal democracy. But an edifice that is built on these alone will not stand; it needs to be buttressed by ensuring the equality of different ethnic groups. For instance, the dominant status of the Punjabis in Pakistan's military and bureaucracy will need to be counterbalanced by empowering the numerically inferior Sindhis, Pushtuns, and Balochis. This is best achieved by granting equal and fair access to state resources. Many of these safeguards are already embodied in the 1973 constitution. Pakistan needs to go back to it as a basic framework and update it in the light of the challenges it is facing today.

But a constitution is only a piece of paper if it is not backed by a social compact between the parties that observe and try to meet its unwritten premises. This is what has been so conspicuously absent in Pakistan. Return to a full-fledged federal arrangement is only the first step towards a progressive building of democratic institutions in Pakistan. A new ethnic contract would mean restructuring Pakistan's federal relations, adding real substance and force to the federal provisions already enacted by the 1973 constitution. It would mean removing formal obstacles to a progressive widening of the governing class and the political base of Pakistan's civilian institutions.

The purpose would be to revive and extend the Grand Bargain in which ethnic and religious communities can exercise power to shape Pakistan. Such a process has three dimensions. The first is to secure an informal but abiding agreement among proximate ethnic communities within a region/province; second, to design an agreement among provinces as coequal partners in governing the unified nation of Pakistan; and third, to establish and institutionalize an agreement between the central state and its parts, the provincial units. There are several federal models available including Pakistan's own experience to draw upon. But the model that might work best for Pakistan is a hybrid model that blends regional and multiethnic federalism in an asymmetrical fashion. Hybrid federalism incorporates unitary features that strengthen the state and allow it to exercise an overarching authority within which it can bargain on behalf of the nation as a whole. The purpose of the overarching authority must be a progressive integration of the Pakistani nation and not the survival of a particular leader or government.

While the granting of provincial autonomy is essential, that alone will not suffice. The federal process also needs to remain open-ended in another respect. The empowerment of one ethnic group will create a succession of similar demands from other ethnic groups and minorities. Every new ethnic mobilization needs to be dealt with in a principled manner. This requires creating a political process that permits representation and accommodation. The dangers of an unresponsive state have been amply visible in India's turbulent northeast, where scores of militant movements compete against ethnically based provincial governments.

There are groups that do not make territorial demands. Pakistan contains a number of smaller ethnic and religious groups that have no clearly defined territorial homeland. Several of these groups have seen their rights severely reduced during the long period of Islamist-backed military rule. Among these are the Christian and Hindu minorities, the Ahmediyas and the Shias.[35] The latter have been the targets of sustained attacks for decades. Unless their status is restored, several key cities like Karachi and Hyderabad will never know real peace. The federal process therefore needs to be revisited periodically. At all times, the central state needs to be not only neutral and transparent but must scrupulously adhere to canons of fairness.

Pakistan can go down this path in a step-by-step manner, dealing with each new demand as it arises. But this could invite charges of political expediency and manipulation and exacerbate conflict. It therefore has a good deal to learn from the strategy of the linguistic reorganization of the Indian states (provinces) in 1957. The states' reorganization was enacted for the county as a whole. It was accepted despite the fact that it denied the claims of separate statehood to several ethnic minorities because the overwhelming majority of ethnic nationalities found the new federal arrangement acceptable and because the criteria upon which statehood was denied or conceded were transparent and impartially applied. The subsequent struggles in India's northeast illustrate why it is important to keep even this federal arrangement open-ended.

The offer of ethno-linguistic autonomy within the framework of a federal Pakistan can become a powerful countermagnet to Islamist nationalism in FATA and NWFP, and even more so in

Balochistan, where the struggle for self-determination is mainly of the older variety. Greater regional autonomy will allow Pakistan to isolate these regions, and the benefits that flow from separating Islamic extremism from ethnic dissidence will benefit the whole country. This is because Pakistan's future as a stable state is premised as much on accommodating grievances in Sindh, Karachi, and Punjab as it is on separating ethnic nationalism from religious extremism.

The process of accommodation has two dimensions: One, the strengthening of existing federal structure, implementing the laws and regulations on the books (the 1973 constitution); and second, informal processes by which a new social compact can be secured with the genuine representatives of ethnic communities. Such a process does not presuppose a full-fledged democracy, but it does require a broad and firm agreement among all the main political actors. By far the most important tacit agreement must revolve around the willingness to accept defeat at the polls and wait for the next round.

Pakistan has already experimented with hybrid federalism of an "illiberal" variety. The most striking example is the different standard of governance applied to FATA. This asymmetrical federalism has become counterproductive because of the spillover effect of war and ethnic irredentism from Afghanistan. If this is to be arrested or rolled back, then an openended, ethnically defined federal bargain needs to be put in place as a counterpoint. This bargain can be asymmetrical so as to accommodate specific histories and cultural traditions or geographical imperatives. Those who would like to see the establishment of a fullfledged democracy with all of its elaborate safeguards for individual human rights may find this less than satisfactory. But for these rights to be given tangible form, it is first necessary to end conflict, and restore stability to Pakistan. That can only be achieved by a progressive expansion of the governing class through a genuine devolution of political power, and not the sham devolution plan that Presidents Zia–ul-Haq and Musharraf had foisted on the country. If this means particular leaders and parties in Pakistan have to lose their lock on power and policy, then they need to accept that outcome. There is no other way to stabilize Pakistan or make its people safe from the scourge of war and violence that has plagued them since the U.S.-Afghan war.

Implication for U.S. Policy

It is clear that nuclear-armed Pakistan—the world's sixth most populous country—has no effective control over a large swath of territory along its border with Afghanistan. Extremist groups that are intent on attacking the United States, such as al Qaeda, enjoy safe haven in these border areas. Recent reports indicate that ISI elements are engaged with groups that support the Taliban and are killing American, NATO, and Afghan troops in Afghanistan.[36] The recent increase in bombings and murders indicate that these terrorist groups have extended their reach into the more settled portions of Pakistan. For most people in Pakistan, the United States is largely to blame for inciting and attacking the Taliban, who had until recently regarded Islamabad as friendly and sympathetic to their cause. According to a recent poll, "only 15 percent of Pakistanis think their country should cooperate with the United States to combat terrorism."[37]

Pakistan's security challenge is compounded by an acute economic crisis. Rising prices and growing violence, the absence of jobs, and poor educational services have pushed the Pakistan youth in search of Jihad. The February 2008 elections ended the 9 years of military rule under Musharraf, but no one is sure whether the new government can rise to meet the challenge and abandon the usual jockeying for office and power.

As a first preparatory step towards stability, Pakistan's leaders and parties need to revive the federal compact, remove all draconian measures and amendments that have been added since Zia's and Musharraf's rule, and devolve power laterally and downward to expand the political base of the central state.

A key prerequisite is the permanent withdrawal of the military from civilian life and political office. Pakistani Chief of Army Staff General Ashfaq Kayani recently stated that the army would stay out of politics. But there is no certainty that the military will honour its promise should the civilian leaders fail to restore calm. The possibility of a future military takeover must be avoided if the United States is serious about its commitment to building a stable Pakistan. The United States needs to take a hard look at its own reasons for backing the military. Many in Washington believe that only a determined and forceful government in Pakistan can eliminate al Qaeda and therefore the United States needs to back the Pakistani military. This may be true, but that is only because in the past the United States has supported military regimes in Pakistan that have perpetuated their stranglehold on society and the economy.

The United States can play a key role in persuading the military to withdraw from politics. Recent literature stresses the "rentier" status of Pakistan, which depends heavily on external capital, especially from the United States, and the periods of accelerated flows of economic assistance (the current period being one of them) coincide with military rule.[38] The United States has given close to $12 billion to Pakistan since 2001, and much of this assistance has been directed to service the needs of the military establishment.[39] According to the recently passed Kerry-Lugar bill, the United States has committed to "empower the Pakistani people charting a path of moderation and stability." The bill is meant to help Pakistan combat al Qaeda and the Taliban by initiating good governance, greater accountability, and respect for human rights. If the civilian authority is strengthened over the military, Pakistan may have a chance to build a stable democracy.[40]

Breaking the self-perpetuating cycle of military rule in Pakistan will take enormous effort and require large scale investment in political and economic assistance. The U.S. Senate has already approved $7 billion for civilian assistance over the next 5 years and $400 million per year for military assistance from 2010-2013, with considerably more in the pipeline. This assistance can come with clearly stated conditions that will urge Pakistan to move towards a genuine federation and lateral power-sharing with its nationalities.

The United States also needs to distinguish between violence caused by the demands of ethnic groups and that caused by Islamic radicalism. Pakistan will require help with integrating the former, while deploying effective counterinsurgency measures against the latter. U.S. assistance should be carefully calibrated to incorporate these different purposes. In addition to linking its long-term commitment to gradual democratization in Pakistan, the United States needs to convince Pakistan to abandon the use of Islamic elements as an instrument of its foreign policy. If this were to happen, India and Pakistan might find their way to settling Kashmir more easily. Peace and moderation can go a long way toward stabilizing Pakistan. The United States can use its considerable influence to persuade Pakistan's leaders to seek both.

Structural changes are usually difficult and most political leaders prefer not to make them for fear of losing control. But a new power-sharing compact among Pakistan's ethnic nationalities will renew the promise of Pakistan and strengthen its central state. It will do this by broadening its base and providing regular channels for resolving ethnic conflict. While several commentators on Pakistan have called for reforms in political parties and elections, improvement in governance, and the accumulation of social capital, Pakistan needs to modify the framework within which these changes can occur. This new framework can be constructed on the basis of a new social compact and a renewed promise to share power to build a stronger Pakistan.

A stable Pakistan is also a Pakistan free from wasteful expenditure on military hardware in an arms race with its traditional enemy, India. The newly forged strategic partnership between India and the United States gives the United States the influence and goodwill to urge New Delhi to take additional confidence-building measures and encourages Pakistan to respond in kind. A friendly India-Pakistan relationship can be a basis to forcefully and vigorously combat the challenges of religious extremism, violence, and poverty in the region. However, the history of the 60-year-old Indo-Pakistani relationship does not inspire confidence that the United States will follow through with these prescriptions. Perhaps with the new Obama administration and a newly elected government in Islamabad, not to mention the return of the Man mohan Singh government in New Delhi, circumstances may be propitious for a recasting of policies all around. There is no doubt that the key to peace and stability and to a terror-free South Asia is a stable and democratic Pakistan.

Endnotes

1. *The Next Chapter, United States and Pakistan*, A Report of the Pakistan Policy Working Group, Washington, DC: United States Institute of Peace, September 2008, *www.usip.org/pubs/ppwg_report.pdf.*
2. For an excellent account of Pakistan's civil military relations, see Hasan-Askari Rizvi, *Military, State and Society in Pakistan*, London, UK: Palgrave Macmillan, 2000.
3. For a brief discussion of these years, see Maya Chadda, *Building Democracy in South Asia*, Boulder, CO: Lynne Rienner, 2000, pp. 67-101.
4. Harsh Pant, ISN Security Watch, *Pakistan: Challenges After Musharraf*, August 22, 2008, available from *www.isn.ethz.ch/isn/*Current-Affairs/Security-Watch/Detail/?ots591=CAB359A3-9328-*19CC-A1D2-8023E646B22C&lng=en&id=90376.*
5. A leading expert on Pakistan imagines scenarios ranging from secular democracy to an Islamic regime along the lines of Iran in Pakistan. See Stephen Cohen, *The Idea of Pakistan*, Washington, DC: Brookings, 2004, pp. 267-301.

6. Daniel Markey, "*Securing Pakistan's Tribal Belt*," Council on Foreign Relations report No. 36, August 2008, available from https://secure.www.cfr.org/content/publications/attachments/ Pakistan_*CSR36.pdf.*
7. Cohen, p. 205.
8. The recent experiment at nation- and state-building in the newly created Bosnia-Herzogovina, itself a fragment of the previous Yugoslavia, underscores the error of dispersing power too widely at the outset.
9. Katherine Adeney, *Federalism and Ethnic Conflict Regulation in India and Pakistan*, New York: Palgrave Macmillan, 2006, p. 141.
10. For a brief discussion, see Chadda, *Building Democracy in South Asia*, pp. 24-30; also see Yogendra Malik, Charles Kennedy, Robert Oberst, Ashok Kapur, Mahendra Lawoti, Syedur Rahman, eds., *Government and Politics in South Asia*, Boulder, CO: Westview Press, 2009, p. 162.
11. Mansoor Akbar Kundi and Arbab Jahangir, "Federalism in Pakistan: Issues and Adjustments," *www.cdrb.org/journal/ 2002/3/2.pdf*; also see Mansoor Akbar Kundi, "Federalism/ Demarcation of Roles for Units in Pakistan," *Asian and African Studies*, 2002, available from *www.cdrb.org/journal/2002/3/ 2.pdf.*
12. International Crisis Group, Asia Report, No. 40, *Pakistan: Transition to Democracy*, October 3, 2002, p. 7.
13. *Ibid.*
14. *Ibid.*, p.4-5
15. For both quotations, see *Ibid.*,
16. Theodore P. Wright, Jr., "Center-Periphery Relations and Ethnic Conflict in Pakistan: Sindhis, Muhajirs, and Punjabis," *Comparative Politics*, Vol. 23, No. 3, April 1991, pp. 301-302.
17. *Ibid.*, p. 302.
18. *Ibid.* As Wright observed, "In the Ayub Khan era, as noted above, the government made considerable land grants to retired army officers and civil servants, both disproportionately

Punjabi, which further exacerbated the conflict, especially if the new landholders were absentees. These settlers were among the first to be attacked in the riots which preceded Ayub's downfall."

19. For both quotations, see Lawrence Ziring cited in Wright, "Center-Periphery Relations," p. 303.
20. "The Exit of Chief Minister," *Dawn Overseas Weekly*, April 14, 1988.
21. Wright, p. 307.
22. Vikhar Ahmed Sayeed, "Muhajirs of Pakistan," *One World South Asia*, August 13, 2008, available from *southasia.oneworld.net/opinioncomment/the-muhajirs-of-pakistan*.
23. Rajat Ganguly, *Kin State Intervention in Ethnic Conflict: Lessons from South Asia*, New Delhi, India: Sage Publications, 1998, pp. 135-136.
24. *Ibid.*, p. 137.
25. Adeel Khan, *Politics of Identity: Ethnic Nationalism and the State in Pakistan*, "New Delhi, India: Sage Publications, 2005, pp. 83-109.
26. P. L. Bhola, "Afridi-Shinwari Revolt in Khyber: Nature, Form, and Intensity," cited in Ganguly, p. 177, note 32.
27. Ganguly, p. 187.
28. Khan, p. 101.
29. There has been a steady stream of bomb and rocket attacks on gas pipelines, railway tracks, power transmission lines, bridges, and communications infrastructure, as well as on military establishments and governmental facilities. Even as the Musharraf government claimed relative success in Balochistan, the more insidious problem of Islamist extremism generated undeniable pressures to respond militarily in NWFP.
30. Christine Fair, Steve Kull, and Clay Ramsey, "Pakistani Public Opinion on Democracy, Islamic Militancy and Relations with the US," A Joint Study by World Public Opinion and the United States Institute for Peace, Washington, DC, February 2008, available from *www.usip.org/pubs/working_papers/wp7_pakistan.pdf*.

31. The TNSM, one of the five outfits proscribed by Musharraf on January 12, 2002, was formed in 1992 with the objective of a militant enforcement of *Sharia*. Ideologically, it is committed to transforming Pakistan into a Taliban-style state. The TNSM operates primarily in the tribal belt, such as Swat and the adjoining districts of the NWFP. Although well established in the NWFP, the TNSM has had only limited success in expanding its activities beyond the tribal areas. It has substantial support in Malakand and Bajaur in the FATA, and includes activists who have fought in Afghanistan at some time during the past 25 years.
32. Kanchan Laxman, "Afflicted Power," *South Asia Intelligence Review, Weekly Review*, Vol. 7, No. 49, June 15, 2009.
33. Seth G. Jones, "Counterinsurgency in Afghanistan," Rand Counterinsurgency Study, Vol. 4, available from *www.rand.org/pubs/monographs/2008/RAND_MG595.pdf.*
34. it is also important to recognize that provision of individual rights is not the only path to a democracy; group rights are equally important, although there is admittedly a tension between the two notions of rights.
35. P.N Khera, "Pakistan Atomised," *Hindustan Times*, December 1, 2006, available from *www.thepersecution.org/news/2006/ht1201.html.*
36. Bruce Riedel, "Pakistan: The Critical Battleground," *Current History*, Vol. 107, No. 712, November 2008, p. 355.
37. Christine Fair, *et al*., "Pakistan Public Opinion," cited in United States Institute for Peace, "The Next Chapter: U.S. and Pakistan," available from *www.usip.org/pubs/ppwg_report.pdf.*
38. Farzana Shaikh, "Pakistan's Perilous Voyage," *Current History*, Vol. 107, No. 712, November 2008, p. 364.
39. *Ibid*.
40. For both quotations, see *American Chronicle*, "Senate Unanimously Passes Kerry-Lugar Pakistan Aid Package," *Congressional Desk*, June 25, 2009, published July 20, 2009.

About the Contributors

HENRY SOKOLSKI is the Executive Director of the Nonproliferation Policy Education Center (NPEC), a Washington, DC-based nonprofit organization founded in 1994 to promote a better understanding of strategic weapons proliferation issues among policymakers, scholars, and the media. He currently serves as an adjunct professor at the Institute of World Politics in Washington, DC, and as a member of the Congressional Commission on the Prevention of Weapons of Mass Destruction Proliferation and Terrorism. Mr. Sokolski previously served as Deputy for Nonproliferation Policy in the Department of Defense, for which he received a medal for outstanding public service from Secretary of Defense Dick Cheney. He also worked in the Office of the Secretary of Defense's Office of Net Assessment, as a consultant to the National Intelligence Council, and as a member of the Central Intelligence Agency's Senior Advisory Group. In the U.S. Senate, he served as a special assistant on nuclear energy matters to Senator Gordon Humphrey (R-NH), and as a legislative military aide to Dan Quayle (R-IN). Mr. Sokolski has authored and edited a number of works on proliferation, including Best of Intentions: America's Campaign Against Strategic *Weapons Proliferation* (Westport, CT: Praeger, 2001); Nuclear Heuristics: Selected Writings of Albert and Roberta *Wohlstetter* (Strategic Studies Institute, 2009); *Falling* Behind: International Scrutiny of the Peaceful Atom (Strategic Studies Institute, 2008); *Pakistan's Nuclear Future: Worries Beyond War* (Strategic Studies Institute, 2008); *Gauging U.S.-Indian Strategic Cooperation* (Strategic Studies Institute, 2007); *Getting*

Ready for a Nuclear-Ready Iran (Strategic Studies Institute, 2005); and *Getting MAD: Nuclear Mutual Assured Destruction, Its Origins and Practice* (Strategic Studies Institute, 2004).

NEIL JOECK is the National Intelligence Officer for South Asia at the National Intelligence Council (NIC). He previously served in the U.S. Government at the National Security Council in the Office of Proliferation Strategy (2004-05) and at the Department of State as a member of the Policy Planning Staff (2001-03). Prior to joining the NIC, he was a senior fellow at the Center for Global Security Research at the Lawrence Livermore National Laboratory (LLNL) and an adjunct professor of political science at the University of California, Berkeley (2005-09). Dr. Joeck worked on India and Pakistan as a political analyst and group leader in Z Division at LLNL (1987-2001) and was a research fellow at the International Institute for Strategic Studies in London (1996-97). Dr. Joeck is the author of *Maintaining Nuclear Stability in South Asia*, Adelphi Paper #312 (1997) and two edited books: *Arms Control and International Security* (with Roman Kolkowicz, 1984) and *Strategic Consequences of Nuclear Proliferation in South Asia* (1986). He has also published numerous journal articles and book chapters. Dr. Joeck holds a BA from the University of California, Santa Cruz; an MA from the Paterson School of International Affairs at Carleton University in Canada; and a Ph.D. and MA in political science from UCLA.

FEROZ HASSAN KHAN is currently on the faculty of the Department of National Security Affairs of the U.S. Naval Postgraduate School, Monterey, CA. He retired as a brigadier general from the Pakistani Army, where he served for 32 years. His previous appointments included the post of Director, Arms Control and Disarmament Affairs within the Strategic Plans Division, Joint Services Headquarters, which is the secretariat of Pakistan's Nuclear Command Authority. His military career blended numerous diplomatic and scholarly assignments. He has experienced combat action and command on active fronts on the line of control in Siachin Glacier and Kashmir. He served domestically and abroad in the United States, Europe, and South

Asia, in particular assisting Pakistan's nuclear diplomacy. General Khan has held a series of visiting fellowships at Stanford University; the Woodrow Wilson International Center for Scholars; the Brookings Institution; the Center for Non-Proliferation Studies at the Monterey Institute of International Studies; and the Cooperative Monitoring Center, Sandia National Laboratory. Since the mid 1990s, General Khan has been making key contributions in formulating and advocating Pakistan's security policy on nuclear and conventional arms control and strategic stability in South Asia. He has produced recommendations for the Ministry of Foreign Affairs and represented Pakistan in several multilateral and bilateral arms control negotiations. He has published and participated in several security related national and international conferences and seminars. He has also been teaching as a visiting faculty member at the Department of the Defense and Strategic Studies, Quaid-e-Azam University, Islamabad. General Khan is currently writing a book on the history of Pakistan's nuclear weapons and U.S. policy; expected publication is in 2010. General Khan holds an M.A. from the Paul Nitze School of Advanced International Studies, The Johns Hopkins University.

JOHN STEPHENSON is a Project Manager at Dalberg Global Development Advisors, a strategy consulting firm focused on international development. He has consulted to the senior management teams of leading international financial institutions, multilateral development organizations, foundations, and multinational corporations on strategy, organizational effectiveness, stakeholder and change management, and development policy. He has experience in several development sectors, including energy and the environment, access to finance, health, private sector development, postconflict reconstruction, and governance & public sector reform. Some of Mr. Stephenson's most recent engagements include: (1) evaluating fund manager proposals and conducting due diligence as part of a $500 million global call for renewable energy funds in emerging markets, (2) serving as a strategic advisor on an innovative $50 million fund for post-conflict countries, (3) working with the United Nations Foundation and Vodafone Group Foundation on their public-private partnership in mobile health and emergency response, and (4) assisting the East African Community to formulate an energy access scale-up strategy to support attainment of the Millennium Development goals with a focus on alternative energy sources. Prior to joining Dalberg, Mr. Stephenson worked at the World Bank where he participated in the formulation of the Bank's Country Assistance Strategy for the Democratic Republic of Congo. He holds a Bachelor's degree magna cum laude in Government and East Asian Studies from Harvard University and a Master's degree from Georgetown University's School of Foreign Service.

PETER TYNAN is a Manager in Dalberg's Washington, DC, office. Mr. Tynan has advised international corporations, governments, and development institutions in strategy, policy, organizational change, supply chain, and performance management. He has experience in several development sectors, including energy, private sector development, public sector reform, and emerging markets investment. Prior to joining Dalberg, Mr. Tynan advised the Minister of Finance of the Democratic Republic of the Congo, where he wrote the private sector revitalization plan and analyzed the competitiveness of major Congolese industries; and worked for the Minister of Finance in Egypt reorganizing the Egyptian Customs Authority. For the U.S. Government, Mr. Tynan advised the CFO of the General Services Administration (GSA) in strategy, strategic planning, and organizational reform. He helped lead the reorganization of the GSA, merging the Supply Service and the Technology Service; and designed and managed the strategic planning process used across the Agency. Mr. Tynan has also worked in private equity, where he sourced and evaluated middle market private equity investments. He is the co-author of *Imagining Australia: Ideas For Our Future* (Allen & Unwin, 2004). Mr. Tynan holds a Bachelor in Business with First Class Honours and the University Medal from the University of Technology in Sydney, Australia; a Masters in Public Policy from the Kennedy School of Government at Harvard University; and an MBA from Harvard Business School.

SHAHID JAVED BURKI, former finance minister of Pakistan, is the chairman of the Institute of Public Policy, a think-tank based in Lahore, Pakistan. He previously served as chief executive officer of EMP Financial Advisors, LLC. He served at the World Bank for 25 years (1974–99), as division chief and senior economist, Policy Planning and Programme Review Department; senior economist and policy adviser, the Office of the Vice President of External Relations; director, International Relations Department of that vice-presidency; director of the China and Mongolia region; and vice president of the Latin American and Caribbean region. Mr. Burki is coauthor of *Sustaining Reform with a US-Pakistan Free Trade Agreement* (2006). His other publications include *Pakistan: Development Choices for the Future* (Oxford University Press, 1986); *Pakistan: Continuing Search for Nationhood* (Westview Press, 1991); *Pakistan: Fifty Years of Nationhood* (Westview Press, 1999); and *Pakistan: A Historical Dictionary* (Scarecrow Press, 1999).

S. AKBAR ZAIDI is one of Pakistan's best known and most prolific political economists. Apart from his interest in political economy, he also has great interest in development, the social sciences more generally, and increasingly, in history. Mr. Zaidi has written over 60 academic articles in international journals and as chapters in books, as well as 11 books and four monographs. His 12th book, *The Political Economy of Democratization in Pakistan*, is forthcoming and the third to be published by Oxford University Press. His other books include *The New Development Paradigm: Papers* on Institutions, NGOs, Gender and Local Government (1999), and *Issues in Pakistan's Economy* (2005), both published by Oxford University Press and both of which are now standard text books for students on Pakistan's economy; and *Pakistan's Economic and Social Development: The Domestic, Regional and Global Context*, (New Delhi, India, Rupa and Co., 2004). He taught at Karachi University for 13 years, and at Johns Hopkins University, where he was a visiting professor in 2004-05. Mr. Zaidi holds a Ph.D. from the University of Cambridge.

CRAIG COHEN is vice president for research and programmes at the Center for Strategic and International Studies (CSIS). He serves as principal liaison between the president's office and research staff, helping the president to connect the growing programme agenda with institutional priorities. Previously, he has served as deputy chief of staff and as a fellow in the Post-Conflict Reconstruction Project of the CSIS International Security Programme. Mr. Cohen was codirector of the CSIS Commission on Smart Power and directed research on Pakistan, authoring *A Perilous Course: U.S. Strategy and Assistance to Pakistan* (CSIS, August 2007). He is also the author of *Measuring Progress in Reconstruction and Stabilization Operations* (United States Institute of Peace, April 2006) and served in 2006 as an adjunct professor at Syracuse University's Maxwell School. Prior to joining CSIS, Mr. Cohen worked with the United Nations and nongovernmental organizations in Rwanda, Azerbaijan, Malawi, and the former Yugoslavia. He received a master's degree from the Fletcher School of Law and Diplomacy at Tufts University and an undergraduate degree from Duke University.

MAYA CHADDA is professor of political science at William Paterson University of New Jersey, and was appointed director of the South Asia Programme at the university in 2004. She has worked for the United Nations Development Programme (UNDP) and the United Nations Family Planning Agency (UNFPA) as a consultant. In 1998, she was appointed the Director of Undergraduate Research for William Paterson University and is currently a Fulbright coordinator at the University. Professor Chadda has served on the review board of the Woodrow Wilson Center and the United States Institute of Peace, Washington, DC, and as a consultant to the John D. and Katherine T. MacArthur Foundation, Minorities Rights Group International (UK), Initiative on Conflict Prevention and Quiet Diplomacy

(Canada). She is a member of the Council on Foreign Relations, where she served on the Joint Task Force of the CFR and Asia Society on South Asia and is currently a member of the American Political Science Association Task Force on "America's Standing in the World." Professor Chadda is the author of *Indo-Soviet Relations* (Bombay, India, Vora & Co.); Paradox of Power: The United States Policy in Southwest *Asia* (Santa Barbara, CA, Clio Press); *Ethnicity Security and Separatism in South Asia* (New York, Columbia University Press/Oxford University Press); and Building Democracy in South Asia: India, Pakistan and *Nepal* (Lynne Rienner/Sage Publishers). Professor Chadda is finishing a contracted manuscript entitled, *Why India Matters*.

Index

NOTES

NOTES

NOTES

NOTES